MST210

Mathematical methods, models and modelling

Book D

This publication forms part of an Open University module. Details of this and other Open University modules can be obtained from the Student Registration and Enquiry Service, The Open University, PO Box 197, Milton Keynes MK7 6BJ, United Kingdom (tel. +44 (0)845 300 6090; email general-enquiries@open.ac.uk).

Alternatively, you may visit the Open University website at www.open.ac.uk where you can learn more about the wide range of modules and packs offered at all levels by The Open University.

To purchase a selection of Open University materials visit www.ouw.co.uk, or contact Open University Worldwide, Walton Hall, Milton Keynes MK7 6AA, United Kingdom for a brochure (tel. +44 (0)1908 858779; fax +44 (0)1908 858787; email ouw-customer-services@open.ac.uk).

Contents

Systems of differential equations

Introduction

Systems of linear first-order differential equations were considered in Unit 6, and for these systems we were able to find an explicit solution. Here we consider qualitative, graphical methods that are applicable to all systems of first-order differential equations, but are of greatest value for those that we cannot solve explicitly, namely non-linear systems.

Recall from Unit 6 that the system of differential equations

$$\dot{x} = -y, \quad \dot{y} = 2x + y,$$

is called *linear* because the right-hand side of each equation does not contain the variables raised to a power (such as x^2) or as the argument of a non-linear function (such as $\sin x$), or products of the variables (such as xy). The systems of differential equations

$$\dot{x} = x^2 + y, \quad \dot{y} = y + t, \tag{1}$$

and

$$\dot{x} = xy, \quad \dot{y} = x^2 y, \tag{2}$$

are both *non-linear* systems of equations.

There is one difference between systems (1) and (2) that is important for this unit. In equations (1) the independent variable t appears explicitly on the right-hand side, whereas in equations (2) only the two dependent variables x and y occur. Systems such as (2), where t does not appear explicitly, are said to be *autonomous*. In this unit we will consider only autonomous equations of the general form

$$\dot{x} = u(x,y), \quad \dot{y} = v(x,y).$$

The methods developed in this unit are widely applicable to a wide variety of situations as differential equation models are common. For this reason we will generally consider the behaviour of systems of differential equations without any specific context, but we do develop one context to illustrate ideas. The situation that we develop is a model of the behaviour of two interacting populations of animals, usually with one variable $x = x(t)$ representing the number of individuals of a predator species, and the other variable $y = y(t)$ representing the number of individuals of its prey.

We use the notations x or $x(t)$, $\dot{x}$ or $\dot{x}(t)$, etc., interchangeably to suit the context.

The graphical methods that we develop use a diagram to give information about solutions of a system without first needing to calculate the solutions. From such diagrams we can answer questions such as deciding whether two populations of animals can coexist with stable populations or whether one population will die out. A solution to the predator and prey model $x(t) = X$, $y(t) = Y$, where X and Y are constants, describes a situation where the two populations coexist with stable populations. Such constant solutions are usually significant, so we give them a name, *equilibrium solutions*, and we call the point (X, Y) in the (x, y)-plane an *equilibrium point*.

Section 1 begins by looking at a way of visualising systems of differential equations and then derives a mathematical model for the interacting population model described above. Section 2 focuses on finding any equilibrium points, and Section 3 looks at classifying the nature of the solutions near an equilibrium point. Section 4 then considers the behaviour of solutions far from equilibrium points.

Note that it is also possible to use a computer to calculate numerical solutions to non-linear equations that cannot be solved explicitly, which is sometimes very useful. However, these numerical solutions are particular solutions with particular initial conditions and cannot answer questions about the set of all solutions in the same way that the graphical methods described in this unit can. These two approaches, graphical and numerical, are in many ways complementary.

1 Visualising systems of differential equations

It is not possible to find algebraic solutions of all systems of differential equations, so we introduce a graphical approach, based on the notion that a point $(x, y) = (x(t), y(t))$ in the plane may be used to represent two variables $x = x(t)$ and $y = y(t)$ at time t. As t increases, the point $(x, y) = (x(t), y(t))$ traces a path that represents the variation of both variables with time. This section also introduces the predator and prey model mentioned in the Introduction, and concludes by describing a method that can be used to derive plenty more examples of systems of differential equations.

1.1 Direction fields revisited

Before considering graphical methods for systems of first-order differential equations, we recall the graphical method for first-order differential equations described in Unit 1, namely direction fields. Consider the differential equation

$$\dot{x} = kx, \quad x > 0, \tag{3}$$

where k is a constant.

This simple differential equation arises in many contexts that involve growth or decay. Among these, it arises as a model for the population size x – which we usually simply refer to as the population x (omitting the word 'size') – as a function of time t. A population x can take only integer values, so we say that x is a discrete variable. It is often convenient to approximate a discrete variable by a variable that can take any real value, referred to as a continuous variable. This approximation will be good if the population size is large. Here we measure populations in hundreds or thousands, as appropriate, so we are able to use quite small numbers to represent large populations in our models. In the continuous model, the derivative $\dot{x}$ represents the rate of increase of the population, which we often refer to as the **growth rate** (even though if $\dot{x} < 0$, it actually represents a decay rate – compare the use in mechanics of 'acceleration' to cover both of the everyday terms 'acceleration' and 'deceleration').

Equation (3) can be solved using the methods of Unit 1, which we ask you to do now.

Exercise 1

Use the methods of Unit 1 to find the particular solution of equation (3) that satisfies the initial condition $x = x_0$ when $t = 0$.

Here we will be concerned not with finding explicit solutions of differential equations such as the one found in Exercise 1, but rather with using graphical methods to determine general features of the solution. As a first step along this path, consider the *direction field* of a differential equation that was introduced in Unit 1.

Figure 1 shows a direction field for $\dot{x} = kx$ together with a graph of the solution that you obtained in Exercise 1. Recall from Unit 1 that the solution is a curve whose tangent at any point has a slope that is equal to the value of the direction field at that point.

The direction field shown at a point indicates the slope of the solution curve passing through that point. Once the direction field is plotted, a sketch of a solution curve can be obtained by smoothly drawing a curve along the lines of the direction field, as shown in Figure 1. This time we calculated the exact solution first, using Exercise 1, but the key point is that we could sketch an approximate solution using the direction field as a guide. In the next section we generalise this from differential equations to systems of differential equations.

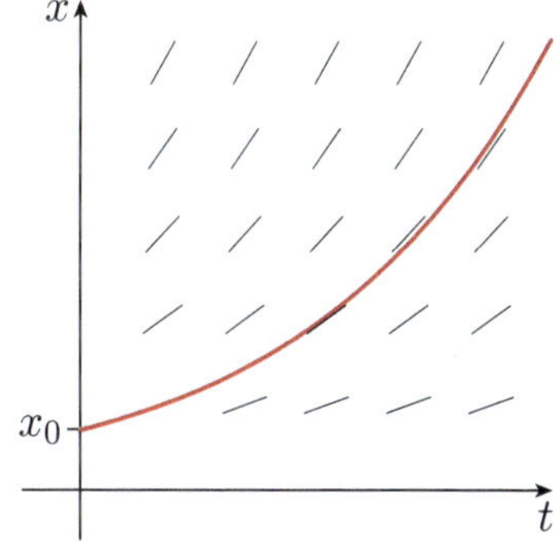

Figure 1 Direction field for $\dot{x} = kx$, with $k > 0$; the red curve shows the solution with initial condition $x(0) = x_0$

1.2 Pictures of solutions

Here we describe a graphical method that can be used to visualise the solutions of systems of first-order differential equations in a similar way to the way in which direction fields can be used to visualise differential equations.

Consider the two variables x and y, which could represent the number of individuals in a pair of populations. Our purpose is to determine how these variables evolve with time. At a particular time t, we suppose that these populations are $x(t)$ and $y(t)$, respectively. We represent this system by a point in the (x, y)-plane. The evolution of the two populations can be represented as a **path**, as shown in Figure 2, where the directions of the arrows on the path indicate the directions in which the point $(x(t), y(t))$ moves along the path with increasing time. Note that this representation does not show how quickly or slowly the point moves along the path but merely the direction of travel.

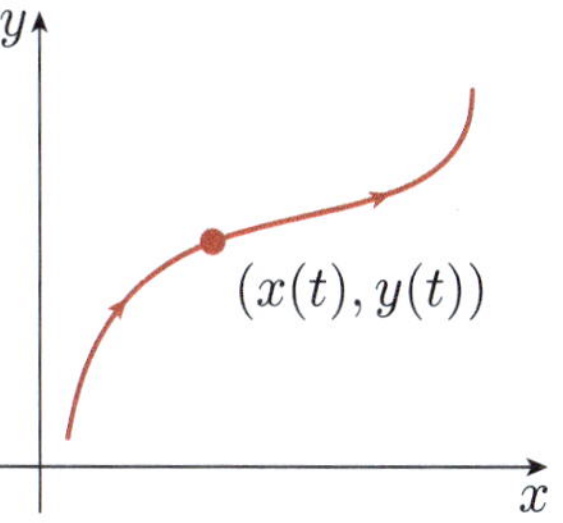

Figure 2 The evolution of two variables is represented by a path in the plane

For the purposes of this discussion, suppose that the populations are evolving independently (perhaps on separate islands) with *no interactions*. The reason for this is that it makes the equations easy to solve, so the solutions can be compared with the graphical method that we are about to introduce. So consider the equations

$$\dot{x} = 0.2x, \quad \dot{y} = 0.3y. \tag{4}$$

(If the variables x and y represent populations, then we must have $x \geq 0$ and $y \geq 0$ in order to be physically reasonable. Here we are interested in developing general methods, so we do not impose this restriction.)

Note that each of equations (4) is in the form of the differential equation (3) considered previously, so they both represent exponential growth.

Equations (4) form a system of linear differential equations, which you met in Unit 6. Using vector notation, the pair of populations may be represented by the vector $\mathbf{x} = (x \quad y)^T$. The system of equations (4) then becomes the vector equation

$$\dot{\mathbf{x}} = \begin{pmatrix} \dot{x} \\ \dot{y} \end{pmatrix} = \begin{pmatrix} 0.2x \\ 0.3y \end{pmatrix}. \tag{5}$$

Vector fields are discussed in more detail in Unit 15.

It is helpful now to introduce the notion of a **vector field**, which is similar to a direction field. In a plane, a direction field associates a direction $f(x, y)$ with each point (x, y), whereas a vector field associates a vector $\mathbf{u}(x, y)$ with each point (x, y). The vector field associated with equation (5) is given by

$$\mathbf{u}(x, y) = \begin{pmatrix} 0.2x \\ 0.3y \end{pmatrix}, \tag{6}$$

so equation (5) becomes $\dot{\mathbf{x}} = \mathbf{u}(x, y)$, which will sometimes be written more simply as $\dot{\mathbf{x}} = \mathbf{u}(\mathbf{x})$. We could use this definition to calculate a direction at any point. For example, at the point $(1, 2)$, the vector field has the direction $\mathbf{u}(1, 2) = (0.2 \times 1 \quad 0.3 \times 2)^T = (0.2 \quad 0.6)^T$. Calculating vectors at several points enables us to construct Figure 3, which shows a plot of this vector field where an arrow in the direction of each vector $\mathbf{u}$ is placed with its midpoint at the point where the vector was calculated.

A direction field $f(x, y)$ represents the slope of a particular solution of the differential equation $dy/dx = f(x, y)$ at the point (x, y). Similarly, $\mathbf{u}(x, y)$ is the vector $\dot{x}\mathbf{i} + \dot{y}\mathbf{j}$ that is tangential to a particular solution of $\dot{\mathbf{x}} = \mathbf{u}(x, y)$ at the point (x, y), because the slope of the tangent is

$$\frac{dy}{dx} = \frac{dy/dt}{dx/dt} = \frac{\dot{y}}{\dot{x}}.$$

This suggests a geometric way of finding a particular solution of equation (5): choose a particular starting point (x_0, y_0), then follow the directions of the tangent vectors. (An exception, which we discuss later, occurs when $\dot{x} = \dot{y} = 0$ at (X, Y), so $\mathbf{u}(X, Y) = \mathbf{0}$ and there is no tangent vector to follow.)

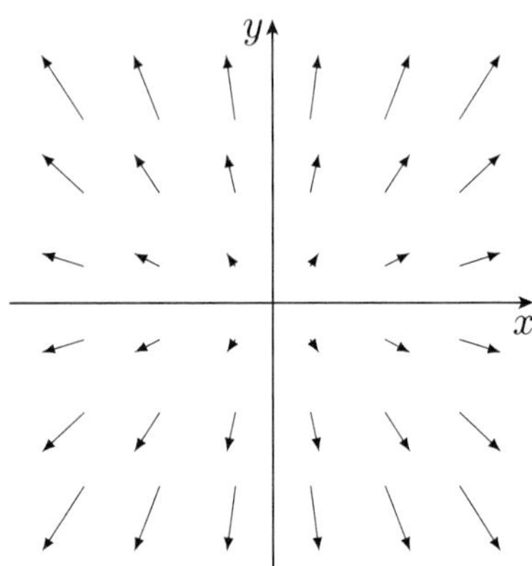

Figure 3 A representation of the vector field $\mathbf{u}(x, y)$

The essential difference between a direction field and a vector field is that the former consists of line segments, and the latter consists of *directed* line segments (which we indicate by arrows) whose lengths indicate the magnitude of $\mathbf{u}(x, y)$. However, since the magnitudes of $\mathbf{u}(x, y)$ may vary considerably and so make the diagram difficult to interpret, we often use arrows of a fixed length to show the direction of a vector field. This is acceptable because, in many cases, it is the direction of $\mathbf{u}(x, y)$ that is our primary concern, rather than its magnitude. Consider Figure 3, where the arrows become longer as distance from the origin increases, so if the arrows are scaled so that the longer arrows do not overlap, then it is hard to discern the direction of the arrows near the origin. To overcome this problem, we use the convention that all arrows will be scaled to the same length. The arrows shown in Figure 4 represent the same vector field but use this scaling convention.

Using the methods of Unit 6, we can find the general solution of equations (4) as

$$x(t) = Ce^{0.2t}, \quad y(t) = De^{0.3t}, \tag{7}$$

where C and D are constants.

This general solution gives the position $(x(t), y(t))$ at time t. It is instructive to plot this solution for a range of time t for various values of the constants C and D. This will show the particular solutions for various initial conditions among the family of general solutions. These particular solutions are overlaid on the vector field plot in Figure 4. The arrows on the solution curves indicate the directions in which the curves are traversed with increasing time.

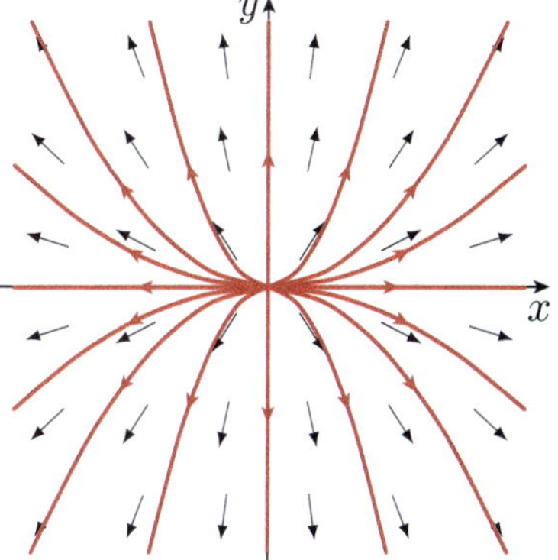

Figure 4   The vector field $\mathbf{u}(x, y)$ together with paths from the family of solutions

A solution curve along which the coordinates x and y vary as t increases is called a **phase path** (or **orbit**). The (x, y)-plane containing the solution curves is called the **phase plane**, and a diagram, such as Figure 4, showing the phase paths is called a **phase portrait**. In other words, a phase portrait is a collection of phase paths that illustrates the behaviour of the differential equations.

You may have noticed in Figure 4 that the paths radiate *outwards* from the origin in all directions. For this reason, we refer to the origin as a **source**.

A source can occur at a point other than the origin.

We now look at the phase paths for a similar system, for which

$$\mathbf{u}(x, y) = \begin{pmatrix} -0.2x \\ -0.3y \end{pmatrix}.$$

Again by using the methods of Unit 6, the general solution can be found as

$$x = Ce^{-0.2t}, \quad y = De^{-0.3t},$$

where C and D are constants.

This solution is in terms of exponential functions with negative exponents, so it represents exponential decay. As this general solution is the same as equations (7) except for the change in sign of the multiples of t, we can see that the paths in the phase plane are the same as before, but with the arrows reversed.

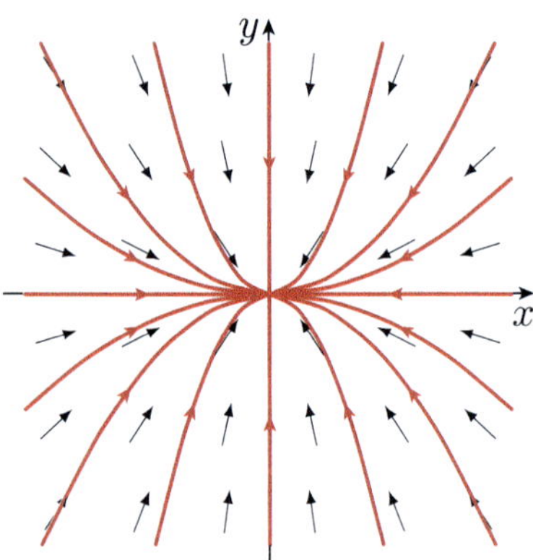

Figure 5 The vector field and solution paths of a *sink*

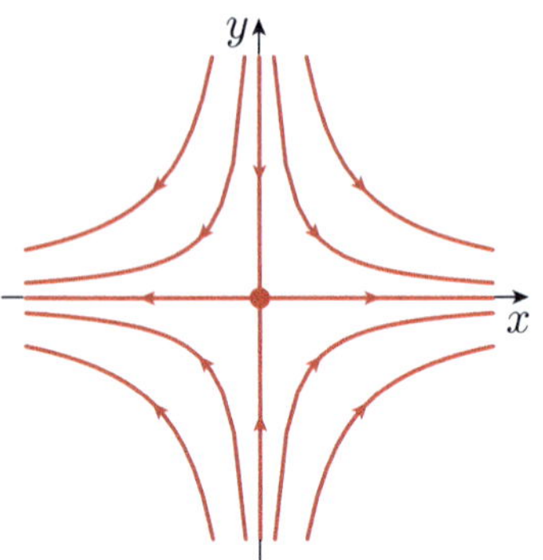

Figure 6 Phase paths for the vector field in Exercise 2, which shows a *saddle point* at the origin

When this model was first proposed, the interaction was between two species of fish in the Adriatic Sea.

Another way of looking at this is to say that at each point in the phase plane, the vector field that represents these equations points in the opposite direction to the previous vector field. This gives the diagram shown in Figure 5. Now the paths radiate inwards *towards* the origin, and for this reason we refer to the origin as a **sink**.

Exercise 2

Write down and solve the system of differential equations $\dot{\mathbf{x}} = \mathbf{u}(x, y)$ given by the vector field

$$\mathbf{u}(x, y) = \begin{pmatrix} x \\ -y \end{pmatrix}.$$

You should find that if you multiply together your solutions for $x(t)$ and $y(t)$ found in Exercise 2, the result is a constant. So the phase paths have the equation $xy = A$, for some constant A. These paths are a family of rectangular hyperbolas (by choosing A to be non-zero) or lines along the axes (by choosing A to be zero). Hence we can sketch the phase portrait for the vector field examined in Exercise 2 to be as shown in Figure 6. You can see that the vast majority of the paths do not radiate into or out of the origin. On these paths, a point initially travels towards the origin, but eventually travels away from it again. The only paths that actually radiate inwards towards or outwards from the origin are those along the x- and y-axes. In this case we call the origin a **saddle point** (the paths look like the contours around a saddle point as defined in Unit 7).

This unit is concerned with sketching phase portraits. Before moving on to give general methods for doing this, we introduce some examples to which we can apply our methods.

1.3 Modelling populations of predators and prey

Here we develop a model for the populations of a predator and its prey. A predator population depends for its survival on there being sufficient prey to provide it with food. Intuition suggests that when the number of predators is low, the prey population may increase quickly, and that this in turn will result in an increase in the predator population. On the other hand, a large number of predators may diminish the prey population until it is very small, and this in turn will lead to a collapse in the predator population. Our mathematical model will need to reflect this behaviour.

To make the discussion more concrete, we will consider modelling populations of foxes and rabbits as the predators and prey. Let $x(t)$ be the number of rabbits, and let $y(t)$ be the number of foxes.

For a population x of rabbits in a fox-free environment, our first model for population change is given by the equation $\dot{x} = kx$, where k is a positive constant. This represents exponential growth. However, if there is a population y of predator foxes, then you would expect the growth rate $\dot{x}$ of rabbits to be reduced.

Similarly, for a population y of foxes in a rabbit-free environment, our first model for the population change is given by the equation $\dot{y} = -hy$, where h is a positive constant. This represents exponential decay. However, if there is a population x of rabbits for the foxes to eat, then you would expect the growth rate $\dot{y}$ of foxes to increase.

In our mathematical model we make the following assumptions.

- There is plenty of vegetation for the rabbits to eat.

- The rabbits are the only source of food for the foxes.

- An encounter between a fox and a rabbit contributes to the fox's larder, which leads directly to a decrease in the rabbit population and indirectly to an increase in the number of foxes.

- The number of encounters between foxes and rabbits is proportional to the number of rabbits multiplied by the number of foxes – that is, it is proportional to xy.

- The growth rate $\dot{x}$ of rabbits decreases by a factor that is proportional to the number of encounters between rabbits and foxes.

- The growth rate $\dot{y}$ of foxes increases by a factor that is proportional to the number of encounters between foxes and rabbits.

These assumptions lead to a differential equation that models the population x of rabbits as

$$\dot{x} = kx - Axy,$$

for some positive constant A. As we will see later, it is convenient to write $A = k/Y$ for some positive constant Y, giving

$$\dot{x} = kx\left(1 - \frac{y}{Y}\right). \tag{8}$$

This is a non-linear equation, since the right-hand side contains an xy term.

Similarly, our revised model for the foxes is given by

$$\dot{y} = -hy + Bxy$$

for some positive constant B. Again, it is convenient to write $B = h/X$ for some positive constant X, so that this equation becomes

$$\dot{y} = -hy\left(1 - \frac{x}{X}\right). \tag{9}$$

Together, the differential equations (8) and (9) model the pair of interacting populations.

This equation can be derived using the input–output principle, which was introduced in Unit 1, Section 1. In a period of time δt, the change in the rabbit population is the number $kx\,\delta t$ of additional rabbits born, taking into account those dying from natural causes, less the number $Axy\,\delta t$ of rabbits eaten.

Again, this equation is non-linear because of the xy term on the right-hand side.

Exercise 3

Sketch the graph of the proportionate growth rate $\dot{x}/x$ of rabbits as a function of the population y of foxes, and the graph of the proportionate growth rate $\dot{y}/y$ of foxes as a function of the population x of rabbits. Interpret these graphs.

Lotka–Volterra equations

The evolution of two interacting populations x and y can be modelled by the **Lotka–Volterra equations**

$$\dot{x} = kx\left(1 - \frac{y}{Y}\right), \quad \dot{y} = -hy\left(1 - \frac{x}{X}\right) \quad (x \geq 0,\ y \geq 0), \tag{10}$$

where x is the population of the prey and y is the population of the predators, and k, h, X and Y are positive constants.

Modelling success

This model was one of the first successful applications of mathematical models to biological systems. It was independently proposed in 1925 by the American biophysicist Alfred Lotka and in 1926 by the Italian mathematician Vito Volterra.

Now we begin to explore the Lotka–Volterra equations.

Exercise 4

(a) Write down the vector field $\mathbf{u}(x, y)$ that corresponds to the Lotka–Volterra equations.

(b) Now suppose that the variables x and y represent thousands of individuals (so that $x = 1$ represents one thousand rabbits, for example) and further suppose that the constants in equations (10) have the values $k = 1$, $h = \frac{1}{2}$, $X = 3$ and $Y = 2$. Complete the following table.

x	y	$\mathbf{u}(x, y)$
0	0	
0	2	
2	0	
2	2	
3	1	
3	2	
3	3	
4	2	

(c) Draw the vectors that you obtained in part (b) as a vector field in the phase plane.

Previously, we were able to solve the pairs of differential equations that arose from our mathematical model, but for equations (10) no explicit formulas for $x(t)$ and $y(t)$ are available. We will use graphical methods to describe the solutions. The vector field plot that you drew in Exercise 4 (partly reproduced as Figure 7(a)) is a starting point. From this plot it may seem possible that the phase path forms a closed loop about the point $(3, 2)$, but this is far from certain. Figure 7(b) shows a sketch of a possible phase path.

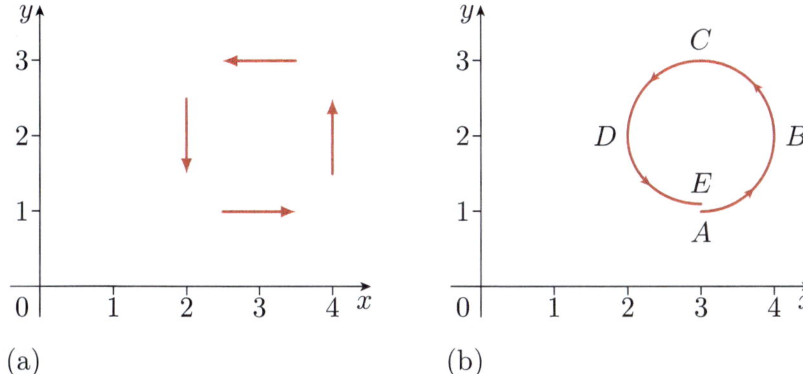

(a)   (b)

Figure 7   (a) Part of the vector field drawn in Exercise 4. (b) A possible path for the changes in rabbit and fox populations that follows the vector field.

Consider a phase path starting at point A, which is the point $(3, 1)$ representing 3000 rabbits and 1000 foxes. From the vector field plot in Exercise 4, we see that the rabbit population increases and so does the fox population, until at the point B we have reached a maximum rabbit population. As the fox population continues to rise, the rabbit population goes into decline. At C, the fox population has reached its maximum, while the rabbits decline further. After this point, there are not enough rabbits available to sustain the number of foxes, and the fox population also goes into decline. At D, the declining fox population gives some relief to the rabbit population, which begins to pick up. Finally, at E, the decline of the fox population is halted as the rabbit population continues to increase. We may even return exactly to the point A, in which case the cycle will repeat indefinitely.

In order to decide whether the cycle will repeat indefinitely, we need to do more than simply plotting a few arrows in the phase plane. This question is discussed further in Section 3 when we have introduced the ideas that are needed to resolve it. Here we conclude this section by looking at a way of generating many more examples of systems of differential equations.

1.4   More examples

In this subsection we describe a method for converting a higher-order differential equation into a system of first-order differential equations. Using this technique we can apply the graphical methods developed in this unit to a much wider range of problems.

As a first example, consider the simple harmonic motion equation

$$\ddot{x} + \omega^2 x = 0, \tag{11}$$

which is a linear second-order differential equation with one dependent variable x and one constant ω. Define $y = \dot{x}$ to be a second variable. Then by differentiation with respect to t we have $\dot{y} = \ddot{x}$. Substituting for $\ddot{x}$ in equation (11) gives $\dot{y} + \omega^2 x = 0$. The two differential equations relating x and y are then

$$\dot{x} = y, \quad \dot{y} = -\omega^2 x,$$

which is a system of two first-order differential equations.

This idea of introducing new variables to represent derivatives of the independent variable can be used to transform any higher-order equation to a system of first-order equations. Try this yourself by attempting the following exercise.

Exercise 5

Convert each of the following differential equations into a system of first-order equations.

(a) $\ddot{x} + \sin x = 0$ (b) $\dfrac{d^2x}{dt^2} - 2\dfrac{dx}{dt} + x^2 = 0$ (c) $\dfrac{d^3u}{dx^3} = 6u\dfrac{du}{dx}$

Using this method you can see that any situation that leads to a differential equation model (such as any dynamics problem) can be converted into a form to which we can apply the graphical methods described in this unit. These methods are widely applicable.

Note that not every system of first-order differential equations can be converted into a single higher-order differential equation by *reversing* the above strategy. For example, the Lotka–Volterra equations cannot be written as a single second-order differential equation in this way.

This strategy of converting a higher-order differential equation to a system of first-order differential equations is also useful in the numerical solution of differential equations. Many numerical methods (e.g. Euler's method, which you met in Unit 1) apply only to first-order differential equations. These methods can be extended in a straightforward manner to apply to systems of first-order equations (by replacing scalar variables with vector variables), and this is the main method for solving higher-order differential equations on a computer. The numerical solutions of the weather prediction model employed to produce daily weather forecasts use exactly this strategy.

Now we have plenty of examples of systems of first-order differential equations and a graphical picture (the phase plane) to represent them. We go on to analyse the features of the graphical picture. First, we look at steady-state solutions of the equations (such as the point $(3, 2)$ in Exercise 4) where two populations can coexist in equilibrium.

2 Equilibrium points

The previous section introduced phase portraits as a method of visualising
systems of non-linear differential equations. Now we begin to investigate
features of phase plane portraits, and we start with the features called
equilibrium points.

2.1 Finding equilibrium points

There is one and only one phase path through a point in the phase plane –
this means that phase paths do not cross. This is because the vector field
has only one direction at each point in the phase plane. But what is
happening at the origin in Figure 8? Here there seem to be phase paths
along the coordinate axes that appear to cross – but all is not as it seems.
The solution along the positive x-axis is given by $x = Ce^t$, $y = 0$, and for
all values of t, the x-coordinate is positive (since the exponential function is
never zero). So the origin is not included on this phase path. Similarly, the
phase paths along the other three axis directions do not include the origin.
The origin is a constant solution of the system of differential equations
that forms a separate path consisting of just a single point. So the phase
paths do not cross at the origin as the origin is on its own path and all the
paths along the coordinate axes approach but never reach the origin.

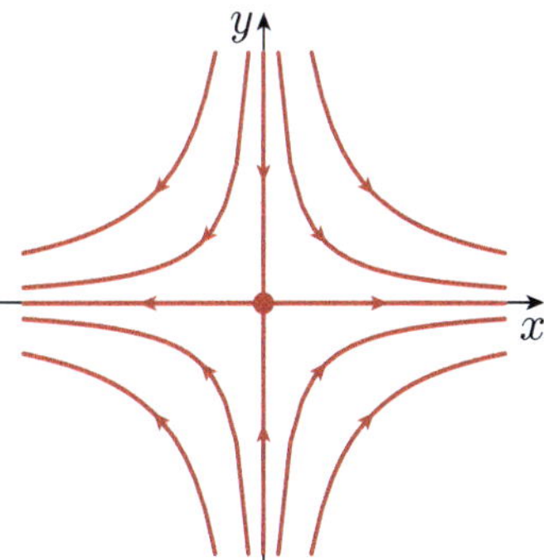

Figure 8   A saddle point at
the origin

These single points, such as the origin in Figure 8, are important features
of phase portraits, so we make the following definition.

An **equilibrium point** of a system of differential equations

$$\dot{\mathbf{x}} = \mathbf{u}(x, y)$$

is a point (X, Y) such that $x(t) = X$, $y(t) = Y$ is a constant solution
of the system, that is, (X, Y) is a point at which $\dot{x}(t) = 0$ and
$\dot{y}(t) = 0$.

Equilibrium points generally represent important physical properties of the
system being modelled. For example, the equilibrium point $(3, 2)$ in
Exercise 4 corresponds to a solution where the populations of rabbits and
foxes are in stable coexistence, that is, they are not changing with time.

The definition of an equilibrium point leads directly to the following
procedure for finding equilibrium points.

> **Procedure 1   Finding equilibrium points**
>
> To find the equilibrium points of the system of differential equations
>
> $$\dot{\mathbf{x}} = \mathbf{u}(x, y)$$
>
> for some vector field $\mathbf{u}$, solve the equation
>
> $$\mathbf{u}(x, y) = \mathbf{0}.$$

Solving $\mathbf{u}(x, y) = \mathbf{0}$ requires the solution of two simultaneous equations, which are generally non-linear, for the unknowns x and y, as the following example shows.

Example 1

Find the equilibrium points for the Lotka–Volterra equations (10) for the rabbit and fox populations. (Remember that h, k, X and Y are *positive* constants.)

Solution

Using Procedure 1, we need to solve the equation $\mathbf{u}(x, y) = \mathbf{0}$, which becomes

$$\begin{pmatrix} kx\left(1 - \dfrac{y}{Y}\right) \\ -hy\left(1 - \dfrac{x}{X}\right) \end{pmatrix} = \begin{pmatrix} 0 \\ 0 \end{pmatrix}.$$

This gives the simultaneous equations

$$kx\left(1 - \frac{y}{Y}\right) = 0, \tag{12}$$

$$-hy\left(1 - \frac{x}{X}\right) = 0. \tag{13}$$

It is important to be methodical when solving simultaneous non-linear equations, in order to avoid missing solutions. Here we will first solve equation (12) and then substitute each solution into equation (13) to find all solutions. Equation (12) is already factorised and the solutions arise when either term is zero, so the solutions are $x = 0$ or $y = Y$.

Substituting $x = 0$ into equation (13) gives $-hy = 0$, so $y = 0$ and hence $(0, 0)$ is an equilibrium point.

Substituting $y = Y$ into equation (13) gives $-hY(1 - x/X) = 0$, so $x = X$ and hence (X, Y) is an equilibrium point. As both X and Y are positive, this equilibrium point is always in the first quadrant (thus is always in the quadrant $x > 0$, $y > 0$ that is physically relevant for population models).

Thus there are two possible equilibrium points for the pair of populations. The first has both the rabbit and fox populations zero, that is, the equilibrium point is at $(0, 0)$; there are no births or deaths – nothing happens. However, the other equilibrium point occurs when there are

X rabbits and Y foxes, that is, the equilibrium point is at (X, Y); the births and deaths exactly cancel out and both populations remain constant.

This explains our choice of constants X and Y in Subsection 1.3.

Now try this yourself by attempting the following exercises.

Exercise 6

Suppose that two variables x and y evolve according to the system of differential equations

$$\dot{x} = x(20 - y), \quad \dot{y} = y(10 - y)(10 - x).$$

Find the equilibrium points of the system.

Exercise 7

Find the equilibrium points of the system of differential equations

$$\dot{x} = 2x^2 y + 7xy^2 + 2y + 1,$$
$$\dot{y} = xy - x.$$

2.2 Stability of equilibrium points

In a real ecosystem it is unlikely that predator and prey populations are in perfect harmony. What if equilibrium is disturbed by a small deviation caused perhaps by a severe winter or hunting? If the number of rabbits is reduced, there would be a decreased food supply for the foxes, and the population of foxes could decrease to zero as a consequence. On the other hand, if the number of foxes is reduced, the birth rate for rabbits would then exceed their death rate, and the number of rabbits could increase without limit.

If a small change or *perturbation* in the populations of rabbits and foxes from their equilibrium values, no matter what the cause, results in subsequent populations that remain close to their equilibrium values, then we say that the equilibrium point is **stable**. On the other hand, if a perturbation results in a catastrophic change, with, for example, the population of foxes or rabbits collapsing to zero or increasing without limit, then we say that the equilibrium point is **unstable**.

In the phase portrait in Figure 9, where the equilibrium point is a sink, you can see that any slight perturbation from the equilibrium point will result in a point that returns to the equilibrium point as time t increases. So this is a *stable* equilibrium point. Similarly, the point $(3, 2)$ in Exercise 4 is a *stable* equilibrium point. In this case a perturbation from the equilibrium point does not result in a point that returns to the equilibrium point as t increases, but it does result in a point that remains in the neighbourhood of the equilibrium point.

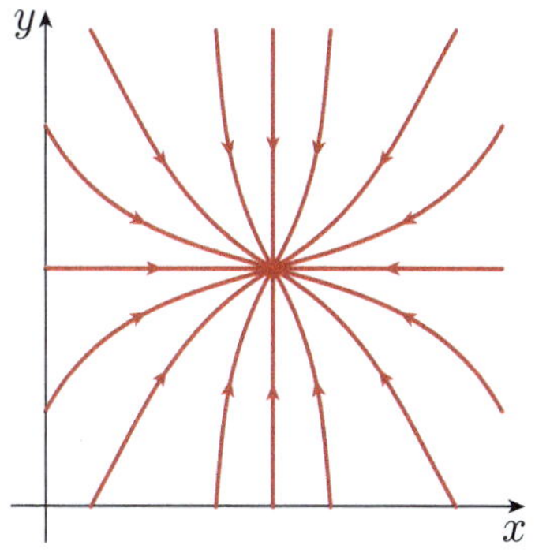

Figure 9   A stable equilibrium point

On the other hand, the origin in the phase portrait of a source shown in Figure 4 is an *unstable* equilibrium point. Any perturbation from the origin will result in the point travelling further and further away from the origin with time. Similarly, the origin in the phase portrait of a saddle shown in Figure 6 is an *unstable* equilibrium point. Apart from increases or decreases in y with x unchanged, any perturbation will result in a point that travels further and further away from the origin with time.

Stability of equilibrium points

An equilibrium point is said to be:

- **stable** when all points in the neighbourhood of the point remain in the neighbourhood of the point as time increases

- **unstable** otherwise.

Exercise 8

Consider the paths shown in Figure 10.

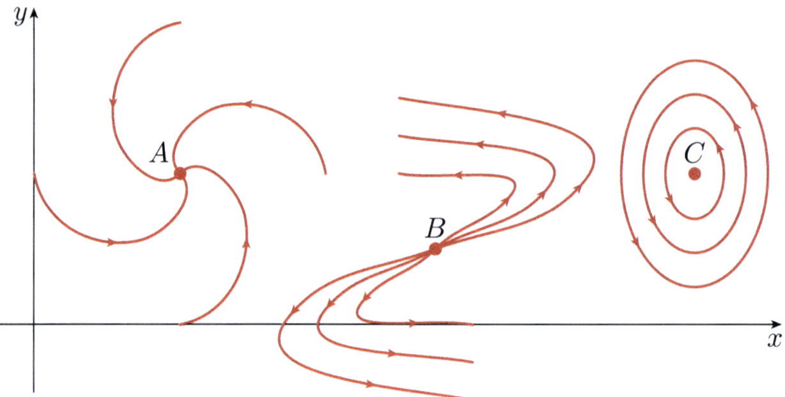

Figure 10   Three equilibrium points labelled A, B and C, together with phase paths in the neighbourhood of each point

Classify as stable or unstable each of the equilibrium points A, B and C.

2.3 Behaviour close to equilibrium

To investigate the behaviour of a non-linear system in the neighbourhood of equilibrium points, in this subsection we develop *linear* approximations to the system that are applicable close to the equilibrium points.

An example will make this clearer, so consider the system of non-linear equations

$$\dot{x} = y(x - 1), \quad \dot{y} = x^2(y - 2). \tag{14}$$

This system of equations has an equilibrium point $(1, 2)$. We wish to investigate the behaviour of solutions close to this equilibrium point, so we change coordinates by letting

$$x = 1 + p, \quad y = 2 + q. \tag{15}$$

For solutions close to equilibrium, (x, y) is close to the point $(1, 2)$, so (p, q) is close to $(0, 0)$, which is another way of saying that p and q should be small. By differentiating equations (15) we obtain $\dot{p} = \dot{x}$ and $\dot{q} = \dot{y}$, so we can write equations (14) as

$$\dot{p} = (2 + q)p = 2p + pq,$$
$$\dot{q} = (1 + p)^2 q = q + 2pq + p^2 q.$$

If p and q are small, then pq is much smaller (and $p^2 q$ is much smaller than this). So neglecting terms that are very small compared to p and q gives the linear system of equations

$$\dot{p} = 2p, \quad \dot{q} = q. \tag{16}$$

As the terms that we neglected were small compared to the terms that we retained, we would expect the solutions of this linear system to approximate the solutions of equations (14) near the equilibrium point. Equations (16) can be solved by separation of variables to give the solutions $p = Ae^{2t}$, $q = Be^t$, where A and B are constants. These equations are exponential growth equations, so the equilibrium point at $p = 0$, $q = 0$ is unstable. The behaviour of the solution for (x, y) close to $(1, 2)$ is just a translation of the behaviour of (p, q) close to $(0, 0)$ thus should be of the same type, so the point $(1, 2)$ is an unstable equilibrium point.

For equations (14) it was straightforward to see which terms to neglect since the right-hand sides of both differential equations were polynomials, but this would not be the case if the right-hand sides involve sinusoidal terms. We now show how Taylor polynomials can be used to find linear approximations in general.

If (X, Y) is an equilibrium point, consider small perturbations p and q giving new populations x and y defined by

$$x = X + p, \quad y = Y + q. \tag{17}$$

We can find the time development of the small perturbations p and q by linearising the differential equation $\dot{\mathbf{x}} = \mathbf{u}(x, y)$. We will make use of Taylor polynomials to achieve this. In order to do so, we must write each component of the vector $\mathbf{u}(x, y)$ as a function of the two variables x and y:

$$\mathbf{u}(x, y) = \begin{pmatrix} u(x, y) \\ v(x, y) \end{pmatrix}.$$

At the equilibrium point (X, Y), we have $\mathbf{u}(X, Y) = \mathbf{0}$, that is,

$$u(X, Y) = 0 \quad \text{and} \quad v(X, Y) = 0.$$

If x and y represent populations, then they cannot be negative. However, the perturbations p and q can (usually) be negative as this would represent populations that are less than the equilibrium values.

Taylor polynomials were introduced in Unit 7.

Now, for small perturbations p and q, we can use the linear Taylor polynomial for functions of two variables to approximate each of $u(x, y)$ and $v(x, y)$ near the equilibrium point (X, Y). Here we use the more compact notation u_x for $\partial u/\partial x$ etc.:

$$u(X + p, Y + q) \simeq u(X, Y) + p\, u_x(X, Y) + q\, u_y(X, Y)$$
$$= p\, u_x(X, Y) + q\, u_y(X, Y),$$

since $u(X, Y) = 0$, and

$$v(X + p, Y + q) \simeq v(X, Y) + p\, v_x(X, Y) + q\, v_y(X, Y)$$
$$= p\, v_x(X, Y) + q\, v_y(X, Y),$$

since $v(X, Y) = 0$.

The above two equations appear rather unwieldy, but are much more succinctly represented in matrix form:

$$\begin{pmatrix} u(x, y) \\ v(x, y) \end{pmatrix} = \begin{pmatrix} u_x(X, Y) & u_y(X, Y) \\ v_x(X, Y) & v_y(X, Y) \end{pmatrix} \begin{pmatrix} p \\ q \end{pmatrix}.$$

Since $x(t) = X + p(t)$ and $y(t) = Y + q(t)$, we also have

$$\dot{x} = \dot{p}, \quad \dot{y} = \dot{q}.$$

Putting the pieces together, substituting in $\dot{\mathbf{x}} = \mathbf{u}(x, y)$ gives a system of *linear* differential equations for the perturbations p and q:

$$\begin{pmatrix} \dot{p} \\ \dot{q} \end{pmatrix} = \begin{pmatrix} u_x(X, Y) & u_y(X, Y) \\ v_x(X, Y) & v_y(X, Y) \end{pmatrix} \begin{pmatrix} p \\ q \end{pmatrix}. \tag{18}$$

Some examples will help to make this clear.

Example 2

Suppose that two variables x and y evolve according to the system of differential equations

$$\dot{x} = x(20 - y), \quad \dot{y} = y(10 - y)(10 - x).$$

The equilibrium points for these equations were found in Exercise 6.

Find the linear approximation to these equations near the equilibrium point $(0, 10)$.

Solution

Here we have

$$u(x, y) = x(20 - y), \quad v(x, y) = y(10 - y)(10 - x).$$

So the partial derivatives are

$$u_x(x, y) = 20 - y, \quad u_y(x, y) = -x,$$
$$v_x(x, y) = -y(10 - y), \quad v_y(x, y) = (10 - y)(10 - x) - y(10 - x).$$

Evaluating these at the given point $(0, 10)$ yields

$$u_x(x, y) = 10, \quad u_y(x, y) = 0,$$
$$v_x(x, y) = 0, \quad v_y(x, y) = -100.$$

So the linear system that approximates the given system near the point $(0, 10)$ is

$$\begin{pmatrix} \dot{p} \\ \dot{q} \end{pmatrix} = \begin{pmatrix} 10 & 0 \\ 0 & -100 \end{pmatrix} \begin{pmatrix} p \\ q \end{pmatrix}.$$

Example 3

Transform the Lotka–Volterra equations (10) into a system of linear differential equations for the perturbations p and q from the equilibrium point (X, Y).

Solution

Here we have

$$u(x, y) = kx\left(1 - \frac{y}{Y}\right), \quad v(x, y) = -hy\left(1 - \frac{x}{X}\right).$$

First, we compute the partial derivatives, obtaining

$$u_x(x, y) = k\left(1 - \frac{y}{Y}\right), \quad u_y(x, y) = -\frac{kx}{Y},$$

$$v_x(x, y) = \frac{hy}{X}, \quad v_y(x, y) = -h\left(1 - \frac{x}{X}\right).$$

Evaluating these at the point (X, Y) gives

$$u_x(X, Y) = 0, \quad u_y(X, Y) = -\frac{kX}{Y},$$

$$v_x(X, Y) = \frac{hY}{X}, \quad v_y(X, Y) = 0.$$

Thus the required system of linear differential equations is

$$\begin{pmatrix} \dot{p} \\ \dot{q} \end{pmatrix} = \begin{pmatrix} 0 & -kX/Y \\ hY/X & 0 \end{pmatrix} \begin{pmatrix} p \\ q \end{pmatrix}. \tag{19}$$

Note that equation (19) can be written as the pair of equations

$$\dot{p} = -\frac{kX}{Y}q, \quad \dot{q} = \frac{hY}{X}p,$$

so we have replaced a system of non-linear equations, for which we have no algebraic solution, with a pair of *linear* equations that we can solve using the methods of Unit 6. We should expect the solutions of equation (19) to provide a good approximation to the original system only when p and q are small (i.e. when the system is close to equilibrium).

Now seems to be an appropriate time to summarise what we have done and to give a name to the matrix that arises. The matrix

$$\mathbf{J}(x, y) = \begin{pmatrix} u_x(x, y) & u_y(x, y) \\ v_x(x, y) & v_y(x, y) \end{pmatrix}$$

is called the **Jacobian matrix** of the vector field

$$\mathbf{u}(x, y) = (u(x, y) \quad v(x, y))^T.$$

The 2×2 matrix on the right-hand side of equation (18) is this Jacobian matrix evaluated at the equilibrium point (X, Y), so equation (18) can be written succinctly as

$$\dot{\mathbf{p}} = \mathbf{J}\mathbf{p},$$

where $\mathbf{p} = (p \quad q)^T$ is the perturbation from the equilibrium point (X, Y), and $\mathbf{J}$ is the Jacobian matrix evaluated at the equilibrium point.

> **Procedure 2   Linearising a system of differential equations near an equilibrium point**
>
> Suppose that the system of differential equations
>
> $$\dot{\mathbf{x}} = \mathbf{u}(x, y) = \begin{pmatrix} u(x, y) \\ v(x, y) \end{pmatrix}$$
>
> has an equilibrium point at $x = X$, $y = Y$.
>
> To linearise this system, carry out the following steps.
>
> 1. Find the Jacobian matrix
>
> $$\mathbf{J}(x, y) = \begin{pmatrix} u_x(x, y) & u_y(x, y) \\ v_x(x, y) & v_y(x, y) \end{pmatrix}. \tag{20}$$
>
> 2. In the neighbourhood of the equilibrium point (X, Y), the differential equations can be approximated by the linearised form
>
> $$\begin{pmatrix} \dot{p} \\ \dot{q} \end{pmatrix} = \mathbf{J} \begin{pmatrix} p \\ q \end{pmatrix}, \tag{21}$$
>
> where $x(t) = X + p(t)$ and $y(t) = Y + q(t)$, and $\mathbf{J} = \mathbf{J}(X, Y)$.

Exercise 9

Write down the linear approximations to the Lotka–Volterra equations (10) near the equilibrium point $(0, 0)$.

Exercise 10

This system was also considered in Exercise 6 and Example 2.

Consider the equations

$$\dot{x} = x(20 - y), \quad \dot{y} = y(10 - y)(10 - x).$$

Find the linear approximations to these equations near the equilibrium point $(10, 20)$.

Exercise 11

Find the equilibrium point of the system of differential equations

$$\dot{x} = 3x + 2y - 8, \quad \dot{y} = x + 4y - 6.$$

Find a system of linear differential equations satisfied by small perturbations p and q from the equilibrium point.

Exercise 12

Suppose that the pair of populations x and y can be modelled by the system of differential equations

$$\dot{x} = 0.5x - 0.00005x^2,$$
$$\dot{y} = -0.1y + 0.0004xy - 0.01y^2$$
$$(x \geq 0, \ y \geq 0).$$

(a) Find the three equilibrium points of the system.

(b) Find the Jacobian matrix of the system.

(c) For each of the three equilibrium points, find the linear differential equations that give the approximate behaviour of the system near the equilibrium point.

We have reduced the discussion of the behaviour of a system near an equilibrium point to an examination of the behaviour of a pair of linear differential equations. In the next section we use the techniques from Unit 6 to solve these differential equations.

3 Classifying equilibrium points

In the previous section you saw how a system of non-linear differential equations $\dot{\mathbf{x}} = \mathbf{u}(x, y)$ may be approximated near an equilibrium point by a linear system $\dot{\mathbf{p}} = \mathbf{J}\mathbf{p}$, where $\mathbf{J}$ is the Jacobian matrix evaluated at the equilibrium point, and $\mathbf{p}$ is a vector of perturbations from the equilibrium point. In this section we develop an algebraic method of classification, based on the eigenvalues of the matrix of coefficients that arises from the linear approximations. Our overall strategy for classifying an equilibrium point of a non-linear system will then be:

- near an equilibrium point, approximate the non-linear system by a linear system

- find the eigenvalues of the matrix of coefficients for this linear approximation

- classify the equilibrium point of the linear system using these eigenvalues

- deduce the behaviour of the original system in the neighbourhood of the equilibrium point.

This section develops the steps of this overall strategy in turn, beginning by looking at the behaviour of systems with Jacobian matrices with two distinct real eigenvalues, complex eigenvalues, and a repeated eigenvalue. These results are then summarised to provide a procedure for classifying equilibrium points of linear systems and then finally non-linear systems.

3.1 Matrices with two distinct real eigenvalues

Let us first consider the system of differential equations $\dot{\mathbf{p}} = \mathbf{J}\mathbf{p}$ where

$$\mathbf{J} = \begin{pmatrix} 2 & 0 \\ 0 & 3 \end{pmatrix}. \tag{22}$$

See Unit 6, Procedure 1.

The matrix $\mathbf{J}$ is diagonal, so the eigenvalues are 2 and 3. The corresponding eigenvectors are $(1 \quad 0)^T$ and $(0 \quad 1)^T$, respectively. The general solution can be written as

$$\begin{pmatrix} p(t) \\ q(t) \end{pmatrix} = C \begin{pmatrix} 1 \\ 0 \end{pmatrix} e^{2t} + D \begin{pmatrix} 0 \\ 1 \end{pmatrix} e^{3t},$$

where C and D are constants, or as

$$p(t) = Ce^{2t}, \quad q(t) = De^{3t}.$$

We are interested in the behaviour of phase paths near the equilibrium point at $p = 0$, $q = 0$. Consider, for example, the paths with $D = 0$ (and $C \neq 0$). On these paths we have $p(t) = Ce^{2t}$ and $q(t) = 0$, so the point $(p(t), q(t))$ moves away from the origin along the p-axis as t increases.

On the other hand, consider the paths with $C = 0$ (and $D \neq 0$). On these paths we have $p(t) = 0$ and $q(t) = De^{3t}$, so the point $(p(t), q(t))$ moves away from the origin along the q-axis as t increases.

Hence we have seen that there are phase paths along the axes, in the directions of the eigenvectors $(1 \quad 0)^T$ and $(0 \quad 1)^T$. A line in the direction of an eigenvector is called an **eigenline**. As t increases, a point on one of the axes moves away from the origin as shown in Figure 11.

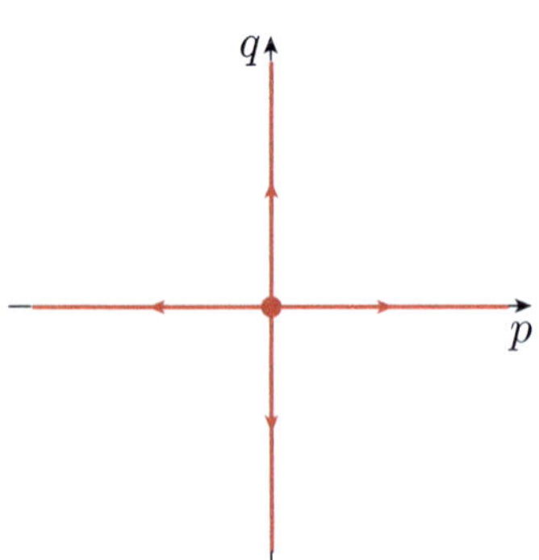

Figure 11 Paths in the (p, q) phase plane; the dot marks the equilibrium point

For general values of C and D, where neither $C = 0$ nor $D = 0$, the point (Ce^{2t}, De^{3t}) still moves away from the origin as t increases, but not along a straight line. As t increases, the point moves along a path that radiates outwards from the origin. An equilibrium point with this type of qualitative behaviour in its neighbourhood is a **source**. This is illustrated in Figure 12, where we have incorporated the fact that the only straight-line paths are along the axes, in the directions of the eigenvectors of the matrix $\mathbf{J}$.

This behaviour occurs for any linear system $\dot{\mathbf{p}} = \mathbf{J}\mathbf{p}$ where the matrix of coefficients $\mathbf{J}$ has *positive distinct eigenvalues*. The only straight-line paths are in the directions of the eigenvectors of the matrix $\mathbf{J}$, as these are the eigenlines (although these will not, in general, be along the axes!).

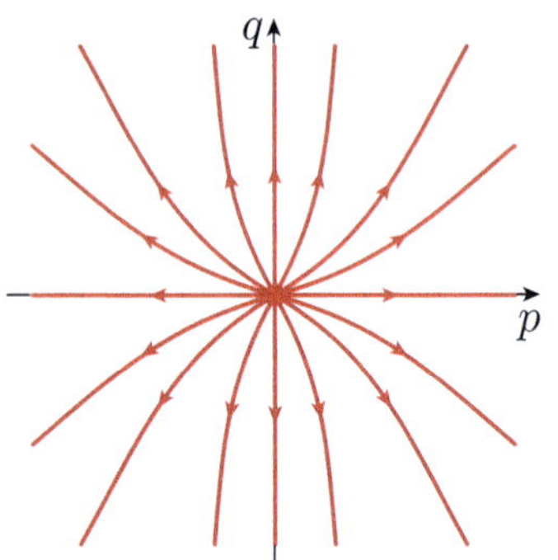

Figure 12 The phase plane near a source

Exercise 13

Consider the linear system of differential equations

$$\begin{pmatrix} \dot{p} \\ \dot{q} \end{pmatrix} = \begin{pmatrix} 3 & 0 \\ 2 & 1 \end{pmatrix} \begin{pmatrix} p \\ q \end{pmatrix}.$$

(a) Find the eigenvalues of the matrix of coefficients.

(b) Classify the equilibrium point $p = 0$, $q = 0$ of the system.

Consider now the system of differential equations $\dot{\mathbf{p}} = \mathbf{J}\mathbf{p}$ where

$$\mathbf{J} = \begin{pmatrix} -2 & 0 \\ 0 & -3 \end{pmatrix}. \tag{23}$$

The change in sign for matrix $\mathbf{J}$ from equation (22) to equation (23) changes the solution from one involving positive exponentials to one involving negative exponentials. You can think of this as replacing t by $-t$, so the solutions describe the same paths, but traversed in opposite directions. A change in the sign of both eigenvalues changes the directions of the arrows along the paths in Figure 12.

If the matrix of coefficients for a linear system has *negative distinct eigenvalues*, the equilibrium point is a **sink**. The only straight-line paths are along the directions of the eigenvectors of the matrix of coefficients.

Consider the linear system of differential equations

$$\begin{pmatrix} \dot{p} \\ \dot{q} \end{pmatrix} = \begin{pmatrix} 0 & -1 \\ 2 & -3 \end{pmatrix} \begin{pmatrix} p \\ q \end{pmatrix}.$$

(a) Find the eigenvalues of the matrix of coefficients.

(b) Classify the equilibrium point $p = 0$, $q = 0$ of the system.

So far in this section we have considered the case where the matrix of coefficients has two distinct positive eigenvalues and the case where the matrix has two distinct negative eigenvalues. We now consider the case where the matrix has *one positive eigenvalue and one negative eigenvalue*. For example, consider the matrix

$$\mathbf{J} = \begin{pmatrix} 1 & 4 \\ 1 & -2 \end{pmatrix},$$

which has eigenvalues 2 and -3, and corresponding eigenvectors $(4 \quad 1)^T$ and $(1 \quad -1)^T$. The general solution of the linear system of differential equations $\dot{\mathbf{p}} = \mathbf{J}\mathbf{p}$ is

$$\begin{pmatrix} p \\ q \end{pmatrix} = C \begin{pmatrix} 4 \\ 1 \end{pmatrix} e^{2t} + D \begin{pmatrix} 1 \\ -1 \end{pmatrix} e^{-3t}, \tag{24}$$

where C and D are constants. When $D = 0$ (and $C \neq 0$), we have $p(t) = 4Ce^{2t}$ and $q(t) = Ce^{2t}$, and the point $(p(t), q(t))$ moves away from the origin along the straight-line path $q = \frac{1}{4}p$ as t increases; this line is an eigenline. On the other hand, when $C = 0$ (and $D \neq 0$), the solution is $p(t) = De^{-3t}$, $q(t) = -De^{-3t}$, so the point $(p(t), q(t))$ approaches the origin along the straight-line path $q = -p$ as t increases.

Hence we have seen that there are two straight-line paths. On the line $q = \frac{1}{4}p$ (which corresponds to the eigenvector $(4 \quad 1)^T$), the point moves away from the origin as t increases. However, on the line $q = -p$ (which corresponds to the eigenvector $(1 \quad -1)^T$), the point moves towards the origin as t increases. These paths are shown in Figure 13.

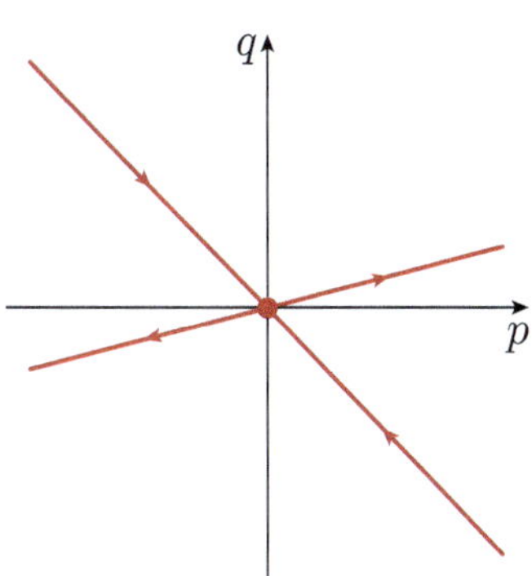

Figure 13   Paths to and from an equilibrium point

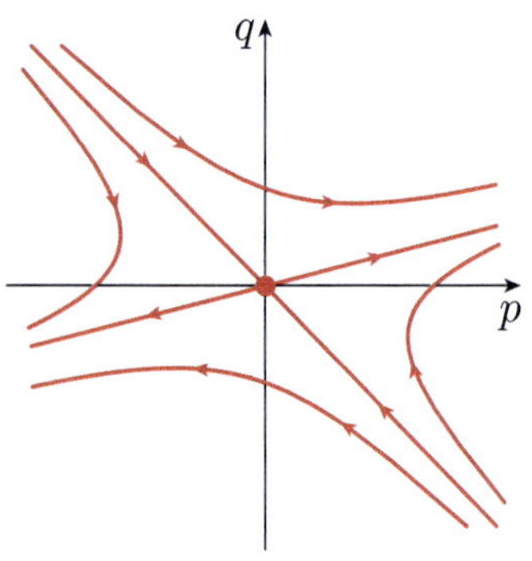

Figure 14   Paths near a *saddle point*; the dot marks the equilibrium point

Now we consider the behaviour of a general point $(p(t), q(t))$, where $p(t)$ and $q(t)$ are given by equation (24), and neither C nor D is zero. For large positive values of t, the terms involving e^{2t} dominate, so $p(t) \simeq 4Ce^{2t}$ and $q(t) \simeq Ce^{2t}$. So for large positive values of t, the general path approaches the line $q = \frac{1}{4}p$. On the other hand, for large negative values of t, the terms involving e^{-3t} dominate, so $p(t) \simeq De^{-3t}$ and $q(t) \simeq -De^{-3t}$. So for large negative values of t, the general path approaches the line $q = -p$. Using this information we can add to Figure 13 to obtain Figure 14.

We can see that the equilibrium point is a **saddle**. The type of behaviour shown in Figure 14 occurs when the matrix of coefficients has *one positive eigenvalue and one negative eigenvalue*. Again, the straight-line paths are in the directions of the eigenvectors of the matrix.

Exercise 15

Consider the linear system of differential equations

$$\begin{pmatrix} \dot{p} \\ \dot{q} \end{pmatrix} = \begin{pmatrix} 1 & 2 \\ 2 & -2 \end{pmatrix} \begin{pmatrix} p \\ q \end{pmatrix}.$$

(a) Find the eigenvalues and corresponding eigenvectors of the matrix of coefficients.

(b) Classify the equilibrium point $p = 0$, $q = 0$.

(c) Sketch the phase paths of the solutions of the differential equations in the neighbourhood of $(0,0)$.

3.2 Matrices with complex eigenvalues

In Unit 5 you saw that some matrices have complex eigenvalues and eigenvectors. In Unit 6 you saw that these complex quantities can be used to construct the *real* solutions of the corresponding system of linear differential equations. Our next example involves such a system.

Example 4

Consider the linear system of differential equations

$$\begin{pmatrix} \dot{p} \\ \dot{q} \end{pmatrix} = \begin{pmatrix} 0 & -1 \\ 4 & 0 \end{pmatrix} \begin{pmatrix} p \\ q \end{pmatrix}.$$

(a) Find the eigenvalues and corresponding eigenvectors of the matrix of coefficients.

(b) Hence write down the general solution of the system of differential equations.

(c) Show that the phase paths for these differential equations are the ellipses

$$p^2 + \tfrac{1}{4}q^2 = K,$$

where K is a positive constant.

Solution

(a) The matrix of coefficients has characteristic equation

$$\begin{vmatrix} -\lambda & -1 \\ 4 & -\lambda \end{vmatrix} = 0,$$

that is, $\lambda^2 + 4 = 0$. So the eigenvalues are $\lambda = 2i$ and $\lambda = -2i$.

When $\lambda = 2i$, the eigenvector $(a \quad b)^T$ satisfies the equation

$$\begin{pmatrix} -2i & -1 \\ 4 & -2i \end{pmatrix} \begin{pmatrix} a \\ b \end{pmatrix} = \begin{pmatrix} 0 \\ 0 \end{pmatrix},$$

so an eigenvector corresponding to the eigenvalue $\lambda = 2i$ is $(1 \quad -2i)^T$.

Similarly, an eigenvector corresponding to the eigenvalue $\lambda = -2i$ is $(1 \quad 2i)^T$.

(b) Using the eigenvalues and corresponding eigenvectors from part (a), and Procedure 3 of Unit 6, the general solution of the differential equations is

$$\begin{pmatrix} p \\ q \end{pmatrix} = C \begin{pmatrix} \cos 2t \\ 2\sin 2t \end{pmatrix} + D \begin{pmatrix} \sin 2t \\ -2\cos 2t \end{pmatrix},$$

where C and D are constants.

(c) We have

$$p(t) = C \cos 2t + D \sin 2t,$$
$$q(t) = 2C \sin 2t - 2D \cos 2t,$$

so

$$\begin{aligned}
p^2 + \tfrac{1}{4}q^2 &= (C \cos 2t + D \sin 2t)^2 + (C \sin 2t - D \cos 2t)^2 \\
&= (C^2 \cos^2 2t + 2CD \cos 2t \sin 2t + D^2 \sin^2 2t) \\
&\quad + (C^2 \sin^2 2t - 2CD \cos 2t \sin 2t + D^2 \cos^2 2t) \\
&= C^2(\cos^2 2t + \sin^2 2t) + D^2(\cos^2 2t + \sin^2 2t) \\
&= C^2 + D^2 = K,
\end{aligned}$$

where $K = C^2 + D^2$ is a positive constant.

So the phase paths are ellipses, as shown in Figure 15. The direction of the arrows can be deduced from the original differential equations. For example, in the first quadrant $\dot{p} < 0$ and $\dot{q} > 0$.

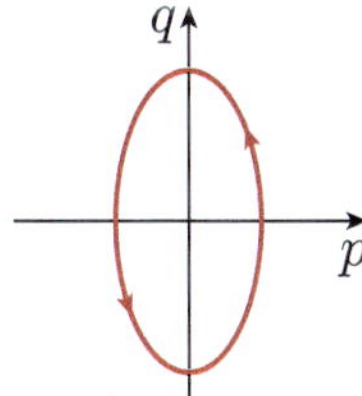

Figure 15   An elliptical path in the phase plane near a *centre*

In Example 4, we saw that all the phase paths are ellipses. This type of behaviour corresponds to any linear system of differential equations where the eigenvalues of the matrix of coefficients are *purely imaginary*. An equilibrium point that has this behaviour in its neighbourhood is called a **centre**.

Exercise 16

Consider the linear system of differential equations

$$\begin{pmatrix} \dot{p} \\ \dot{q} \end{pmatrix} = \begin{pmatrix} 2 & -1 \\ 5 & -2 \end{pmatrix} \begin{pmatrix} p \\ q \end{pmatrix}.$$

(a) Find the eigenvalues of the matrix of coefficients.

(b) Classify the equilibrium point $p = 0$, $q = 0$.

In general, when the eigenvalues of a matrix are complex, they are not purely imaginary but also contain a real part. This has a significant effect on the solution of the corresponding system, as you will see in the following example.

Example 5

Find the general solution of the system of equations $\dot{\mathbf{p}} = \mathbf{J}\mathbf{p}$, where

$$\mathbf{J} = \begin{pmatrix} -2 & -3 \\ 3 & -2 \end{pmatrix}.$$

Sketch some paths corresponding to the solutions of the system.

Solution

The characteristic equation of the matrix of coefficients is $(2 + \lambda)^2 + 9 = 0$, so the eigenvalues are $-2 + 3i$ and $-2 - 3i$. Corresponding eigenvectors are $(1 \quad -i)^T$ and $(1 \quad i)^T$, respectively, so the general solution is given by

$$\begin{pmatrix} p \\ q \end{pmatrix} = Ce^{-2t} \begin{pmatrix} \cos 3t \\ \sin 3t \end{pmatrix} + De^{-2t} \begin{pmatrix} \sin 3t \\ -\cos 3t \end{pmatrix},$$

where C and D are constants.

If we neglect, for the time being, the e^{-2t} terms, the solution is

$$p = C \cos 3t + D \sin 3t,$$
$$q = C \sin 3t - D \cos 3t,$$

from which it follows that

$$p^2 + q^2 = C^2 + D^2.$$

So in the absence of the e^{-2t} terms, the paths would be circles with centre at the origin. The effect of the e^{-2t} terms on these paths is to reduce the radius of the circles gradually. In other words, the paths spiral in towards the origin as t increases, as shown in Figure 16.

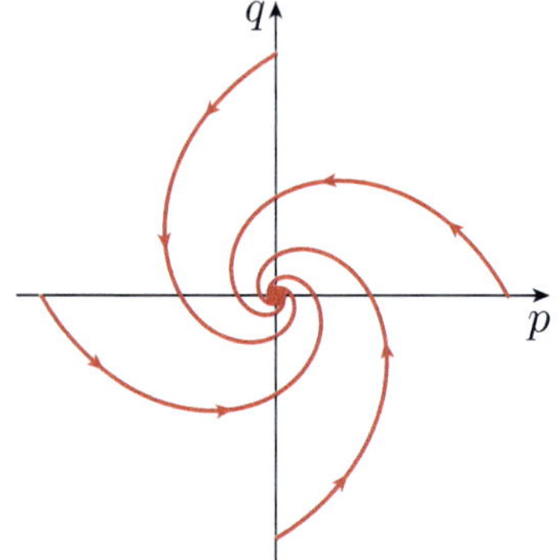

Figure 16 Paths near a *spiral sink*; the dot marks the equilibrium point

In Example 5 (and Figure 16) the paths spiral in towards the origin, so the origin is a sink (called a **spiral sink**) and therefore is a *stable* equilibrium point.

If the paths spiralled away from the origin, we should have a
spiral source (see Figure 17) with the equilibrium point *unstable*.

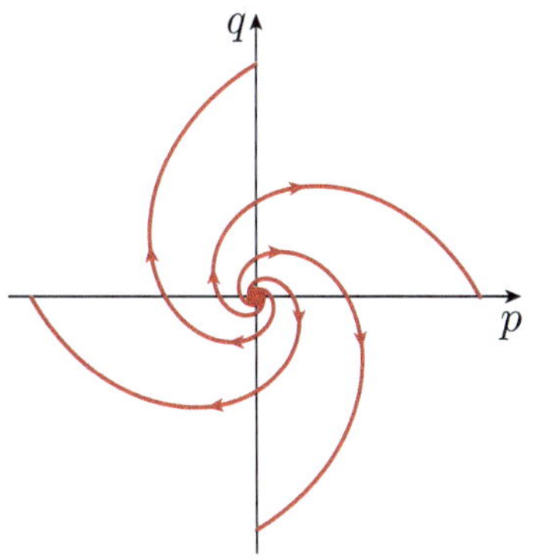

The stability is determined by the sign of the real part of the complex
eigenvalues. To summarise, if the real part is positive, then the general
solution involves e^{kt} terms (where k is positive) and the equilibrium point
is a spiral source; if the real part is negative, then the general solution
involves e^{-kt} terms and the equilibrium point is a spiral sink.

Figure 17   Paths near a
spiral source; the dot marks
the equilibrium point

Exercise 17

Consider the linear system of differential equations

$$\begin{pmatrix} \dot{p} \\ \dot{q} \end{pmatrix} = \begin{pmatrix} 1 & 1 \\ -1 & 1 \end{pmatrix} \begin{pmatrix} p \\ q \end{pmatrix}.$$

(a) Find the eigenvalues of the matrix of coefficients.

(b) Classify the equilibrium point $p = 0$, $q = 0$.

3.3 Matrices with repeated eigenvalues

So far we have considered the cases where the matrix of coefficients for a
linear system of differential equations has two real distinct eigenvalues or
complex eigenvalues. In this subsection we consider a third possibility: the
case where the matrix has a real repeated eigenvalue. In fact, there are two
separate cases, depending on how many independent eigenvectors there are.

First, we consider the case where there are two linearly independent
eigenvectors. In this case, the matrix must be diagonal and of the form

$$\mathbf{J} = \begin{pmatrix} \lambda & 0 \\ 0 & \lambda \end{pmatrix}$$

(see Unit 5, Exercise 13(a)).

Exercise 18

Consider the linear system of differential equations

$$\begin{pmatrix} \dot{p} \\ \dot{q} \end{pmatrix} = \begin{pmatrix} 2 & 0 \\ 0 & 2 \end{pmatrix} \begin{pmatrix} p \\ q \end{pmatrix}.$$

(a) Find the eigenvalues and eigenvectors of the coefficient matrix.

(b) Find the general solution of the system of differential equations.

(c) By eliminating t, find the equations of the paths, and describe them.

(d) Is the equilibrium point $p = 0$, $q = 0$ stable or unstable?

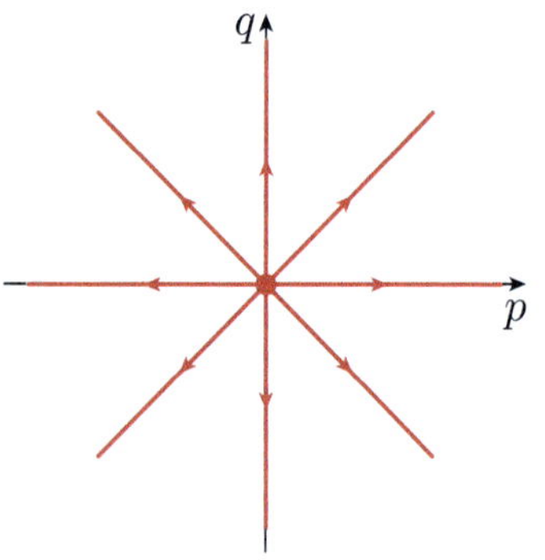

Figure 18 Paths near a *star source*; the dot marks the equilibrium point

In Exercise 18, we have seen that when the matrix of coefficients has two real *identical positive eigenvalues* and *two linearly independent eigenvectors*, all the paths are straight lines radiating away from the origin, as shown in Figure 18. The equilibrium point at $p = 0$, $q = 0$ is called a **star source**. If there are *two identical negative eigenvalues* (but still *two linearly independent eigenvectors*), then the arrows on the paths in Figure 18 are reversed, and the equilibrium point at $p = 0$, $q = 0$ is called a **star sink**.

We turn finally to the case where there are two identical eigenvalues but only one independent eigenvector.

Exercise 19

(a) Find the eigenvalues and corresponding eigenvectors of the matrix
$$\mathbf{J} = \begin{pmatrix} 2 & 0 \\ 1 & 2 \end{pmatrix}.$$

(b) Find the general solution of the system of differential equations
$$\dot{\mathbf{p}} = \mathbf{J}\mathbf{p}.$$

In Exercise 19 we have seen that the general solution of the system of linear differential equations
$$\begin{pmatrix} \dot{p} \\ \dot{q} \end{pmatrix} = \begin{pmatrix} 2 & 0 \\ 1 & 2 \end{pmatrix} \begin{pmatrix} p \\ q \end{pmatrix}$$

is
$$p(t) = Ce^{2t}, \quad q(t) = (Ct + D)e^{2t},$$

where C and D are constants.

Figure 19 shows some typical paths of this system, and we conclude that the equilibrium point at $p = 0$, $q = 0$ is unstable. It is called an **improper source**. Observe that there is only one straight-line path leading away from the equilibrium point (the y-axis in this case), which is a consequence of the fact that the Jacobian matrix has only one eigenvector. If the (repeated) eigenvalue is *negative*, then the phase paths are obtained by reversing the arrows in Figure 19 to obtain an **improper sink**.

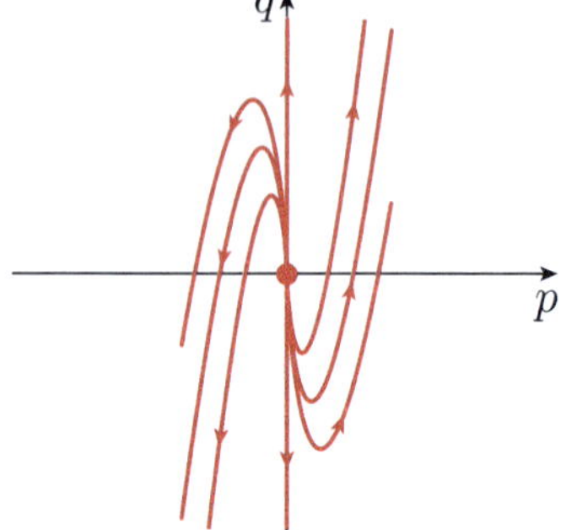

Figure 19 Paths near an *improper source*; the dot marks the equilibrium point

This completes the discussion of our range of examples, which were chosen to illustrate most types of behaviour that you might meet. These are summarised in the next subsection.

3.4 Classifying equilibrium points of linear systems

In Procedure 3 we summarise the results of the previous three subsections. Note that this procedure is not exhaustive because it does not include a number of degenerate cases where one of the eigenvalues is zero, which are not considered in this module.

Procedure 3   Equilibrium point classification for a linear system

Consider the linear system $\dot{\mathbf{p}} = \mathbf{J}\mathbf{p}$ for a 2×2 matrix $\mathbf{J}$. The nature of the equilibrium point at $p = 0$, $q = 0$ is determined by the eigenvalues and eigenvectors of $\mathbf{J}$, and a decision tree for this is shown in Figure 20.

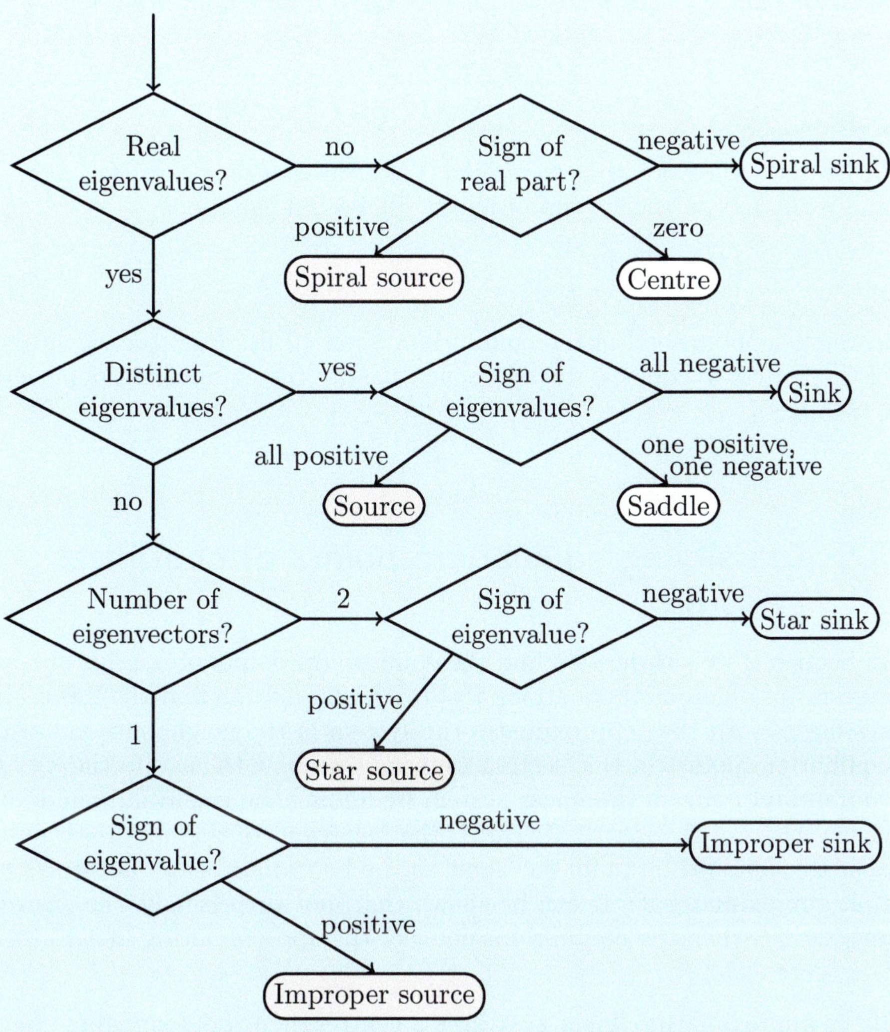

Figure 20   A decision tree for classifying equilibrium points

---

Exercise 20

In Example 3 we saw that the Lotka–Volterra equations can be approximated by the system of linear differential equations

$$\begin{pmatrix} \dot{p} \\ \dot{q} \end{pmatrix} = \begin{pmatrix} 0 & -kX/Y \\ hY/X & 0 \end{pmatrix} \begin{pmatrix} p \\ q \end{pmatrix}$$

in the neighbourhood of the equilibrium point (X, Y). Find the eigenvalues of the matrix of coefficients, and hence classify the equilibrium point $p = 0$, $q = 0$.

Exercise 21

In Exercise 9 we saw that the Lotka–Volterra equations can be approximated by the system of linear differential equations

$$\begin{pmatrix} \dot{p} \\ \dot{q} \end{pmatrix} = \begin{pmatrix} k & 0 \\ 0 & -h \end{pmatrix} \begin{pmatrix} p \\ q \end{pmatrix}$$

in the neighbourhood of the equilibrium point $(0, 0)$. Find the eigenvalues of the matrix of coefficients, and hence classify the equilibrium point $p = 0$, $q = 0$.

3.5 Classifying equilibrium points of non-linear systems

In Section 2 we saw how to find the equilibrium points of non-linear systems of differential equations $\dot{\mathbf{x}} = \mathbf{u}(\mathbf{x})$, and how to find the linear system $\dot{\mathbf{p}} = \mathbf{J}\mathbf{p}$ that approximates the system in the neighbourhood of the equilibrium point. In this section we have seen how to classify the equilibrium point of the linear system by finding the eigenvalues and eigenvectors of the matrix $\mathbf{J}$. But is the behaviour of the non-linear system near the equilibrium point the same as the behaviour of the linear system that approximates it? It can be shown that, not surprisingly, the answer is yes, *except* when the equilibrium point of the approximating linear system is a centre.

If the approximating linear system is a centre, then the dominant behaviour of the original system is to circulate around the equilibrium point. Consider what happens when we follow one circuit of a path around the equilibrium point of a non-linear system, such as that shown in Figure 21. This figure shows the equilibrium point at point P and a path $ABCDE$ that circles around it. After one circuit, we compare the distance from the equilibrium point EP with the starting distance AP.

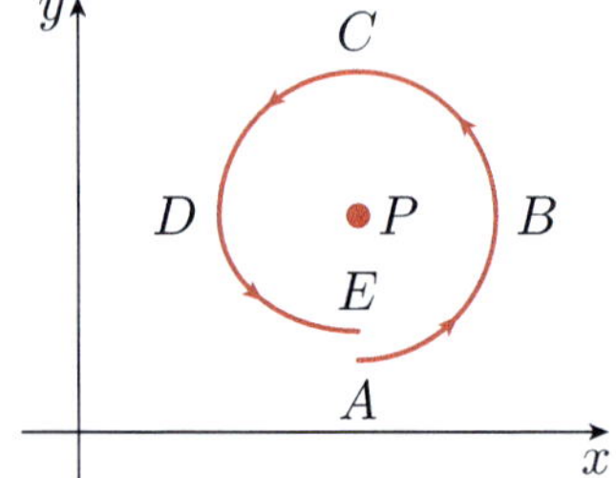

Figure 21 One circuit around an equilibrium point P of a non-linear system where the linearised system is a centre

There are three possibilities: $AP < EP$, $AP = EP$ and $AP > EP$ (the latter is shown in the figure). These three possibilities correspond to a spiral source, a centre or a spiral sink, respectively. We now give examples to show that all three possibilities can occur.

A very simple example of a system with a centre at the origin is the linear system

$$\dot{x} = -y, \quad \dot{y} = x. \tag{25}$$

To see that the origin is a centre, differentiate the second equation with respect to t to get $\ddot{y} = \dot{x}$, and use the first equation to substitute for $\dot{x}$ to obtain $\ddot{y} = -y$. You should recognise this equation as the simple harmonic motion equation, which has solution $y = A\cos(t + \phi)$ where A and ϕ are constants. Hence we have $x = \dot{y} = -A\sin(t + \phi)$. So the general solution of the system of differential equations is $(x, y) = (-A\sin(t + \phi), A\cos(t + \phi))$, which is the family of concentric circles $x^2 + y^2 = A^2$.

There are other non-linear systems of equations that have equations (25) as a linear approximation, for example all equations using these linear terms with some higher powers of x or y added. Consider one such example:

$$\dot{x} = -y, \quad \dot{y} = x + y^3. \tag{26}$$

If we proceed as before and differentiate the second equation and substitute for $\dot{x}$, then we obtain $\ddot{y} - 3y^2\dot{y} + y = 0$. This non-linear equation is harder to solve, but we can get an idea of the solution by approximating the equation. Since y^2 is always positive, we can get a feel for the solutions by looking at the solutions of the equation $\ddot{y} - k\dot{y} + y = 0$, where k is a positive constant. The constant k is small for paths close to the equilibrium point, so the solutions of the characteristic equation are complex with positive real part, which implies that the solutions of the linear second-order equation increase exponentially. So we expect the solutions of equations (26) to also spiral outwards (as can be confirmed by plotting the solution on a computer). So (26) is an example of a non-linear system that is a spiral source where the linear approximation (equations (25)) has a centre.

Changing the sign in front of the y^3 term gives a non-linear system that has a spiral sink at the origin with the same linear approximation:

$$\dot{x} = -y, \quad \dot{y} = x - y^3.$$

Plotting the solutions of this system on a computer confirms that this is a non-linear system that has a spiral sink at the origin.

Thus if the linear approximation has a centre, we cannot immediately deduce the nature of the equilibrium point of the original non-linear system: it may be a centre, a spiral sink or a spiral source. This is summarised in the following procedure.

Procedure 4   Equilibrium point classification for a non-linear system

To classify the equilibrium points of a system of non-linear differential equations, carry out the following steps.

1. Find the equilibrium points by using Procedure 1.

2. Use Procedure 2 to find a linear system that approximates the original non-linear system in the neighbourhood of each equilibrium point.

3. For each equilibrium point, use Procedure 3 to classify the linear system.

4. For each equilibrium point, the behaviour of the original non-linear system is the same as that of the linear approximation, except when the linear system has a centre. If the linear system has a centre, then the equilibrium point of the original non-linear system may be a centre, a spiral sink or a spiral source.

The following example shows how this procedure is used.

Example 6

Consider the non-linear system of differential equations

$$\dot{x} = -4y + 2xy - 8, \quad \dot{y} = 4y^2 - x^2.$$

(a) Find the equilibrium points of the system.

(b) Compute the Jacobian matrix of the system.

(c) In the neighbourhood of each equilibrium point:

- linearise the system of differential equations

- classify the equilibrium point of the linearised system.

(d) What can you say about the classification of the equilibrium points of the original (non-linear) system of differential equations?

Solution

(a) The equilibrium points are given by

$$-4y + 2xy - 8 = 0,$$
$$4y^2 - x^2 = 0.$$

The second equation gives

$$x = \pm 2y.$$

When $x = 2y$, substitution into the first equation gives

$$-4y + 4y^2 - 8 = 0,$$

or $y^2 - y - 2 = 0$, which factorises to give

$$(y - 2)(y + 1) = 0.$$

Hence

$$y = 2 \quad \text{or} \quad y = -1.$$

When $y = 2$, $x = 2y = 4$. When $y = -1$, $x = 2y = -2$. So we have found two equilibrium points, namely $(4, 2)$ and $(-2, -1)$.

When $x = -2y$, substitution into the first equation gives

$$-4y - 4y^2 - 8 = 0,$$

or $y^2 + y + 2 = 0$. This quadratic equation has no real solutions, so there are no more equilibrium points.

(b) Differentiating the right-hand sides of the given differential equations gives the Jacobian matrix as

$$\mathbf{J} = \begin{pmatrix} 2y & 2x - 4 \\ -2x & 8y \end{pmatrix}.$$

(c) At the equilibrium point $(4, 2)$, the Jacobian matrix is

$$\begin{pmatrix} 4 & 4 \\ -8 & 16 \end{pmatrix},$$

so the linearised system is

$$\begin{pmatrix} \dot{p} \\ \dot{q} \end{pmatrix} = \begin{pmatrix} 4 & 4 \\ -8 & 16 \end{pmatrix} \begin{pmatrix} p \\ q \end{pmatrix}.$$

The characteristic equation of the matrix of coefficients is

$$(4 - \lambda)(16 - \lambda) + 32 = 0,$$

or $\lambda^2 - 20\lambda + 96 = 0$, which factorises to give

$$(\lambda - 8)(\lambda - 12) = 0,$$

so the eigenvalues are

$$\lambda = 8 \quad \text{and} \quad \lambda = 12.$$

The two eigenvalues are positive and distinct, so the equilibrium point $p = 0$, $q = 0$ is a source.

At the equilibrium point $(-2, -1)$, the Jacobian matrix is

$$\begin{pmatrix} -2 & -8 \\ 4 & -8 \end{pmatrix},$$

so the linearised system is

$$\begin{pmatrix} \dot{p} \\ \dot{q} \end{pmatrix} = \begin{pmatrix} -2 & -8 \\ 4 & -8 \end{pmatrix} \begin{pmatrix} p \\ q \end{pmatrix}.$$

The characteristic equation of the matrix of coefficients is

$$(-2 - \lambda)(-8 - \lambda) + 32 = 0,$$

which simplifies to

$$\lambda^2 + 10\lambda + 48 = 0.$$

The roots of this quadratic equation are

$$\lambda = -5 \pm i\sqrt{23},$$

so the eigenvalues are complex with negative real part. Hence the equilibrium point $p = 0$, $q = 0$ is a spiral sink.

(d) As neither of the equilibrium points in part (c) is a centre, the non-linear system has an equilibrium point $(4, 2)$ that is a source, and an equilibrium point $(-2, -1)$ that is a spiral sink.

The following exercises ask you to follow Procedure 4 to classify some equilibrium points.

Exercise 22

A certain system of differential equations has an equilibrium point where the Jacobian matrix evaluates to

$$\mathbf{J} = \begin{pmatrix} 2 & -3 \\ 3 & 2 \end{pmatrix}.$$

(a) Find the eigenvalues and eigenvectors of this matrix, and hence classify the equilibrium point.

(b) If this Jacobian matrix is a linear approximation to a non-linear system $\dot{\mathbf{x}} = \mathbf{u}(x, y)$, what can you say about the equilibrium point of the non-linear system?

Exercise 23

Consider the non-linear system of differential equations

$$\dot{x} = (1 + x - 2y)x, \quad \dot{y} = (x - 1)y.$$

(a) Find the equilibrium points of the system.

(b) Find the Jacobian matrix of the system.

(c) In the neighbourhood of each equilibrium point:

- find the linear system of differential equations that gives the approximate behaviour of the system near the equilibrium point

- find the eigenvalues of the matrix of coefficients

- use the eigenvalues to classify the equilibrium point of the linearised system.

(d) What can you say about the classification of the equilibrium points of the original non-linear system of differential equations?

This section concludes with a brief discussion of the nature of the equilibrium point (X, Y) of the Lotka–Volterra equations (10). Near this equilibrium point, the linear system of differential equations that approximates the Lotka–Volterra equations has a centre (as you saw

in Exercise 20). So using Procedure 4, the equilibrium point of the
Lotka–Volterra equations is a centre, a spiral sink or a spiral source.

Methods for deciding between these three options are beyond the scope of
this module, but for completeness we outline an argument for the
Lotka–Volterra equations. The key here is to find a relationship between x
and y that the phase paths satisfy, in the same way that we showed that
the paths were ellipses in Example 4(c). For the Lotka–Volterra equations,
we consider the function

$$f(x,y) = h \ln x - \frac{h}{X}x + k \ln y - \frac{k}{Y}y \quad (x > 0,\ y > 0).$$

How this function arises is explained below; all we are doing here is starting from this given function.

The key fact about this particular function f is that it is constant along
the phase paths. To show this, differentiate with respect to t:

$$\frac{df}{dt} = \frac{dx}{dt}\frac{d}{dx}\left(h \ln x - \frac{h}{X}x\right) + \frac{dy}{dt}\frac{d}{dy}\left(k \ln y - \frac{k}{Y}y\right)$$

$$= \dot{x}\left(\frac{h}{x} - \frac{h}{X}\right) + \dot{y}\left(\frac{k}{y} - \frac{k}{Y}\right)$$

$$= \dot{x}\frac{h}{x}\left(1 - \frac{x}{X}\right) + \dot{y}\frac{k}{y}\left(1 - \frac{y}{Y}\right).$$

Now substitute for $\dot{x}$ and $\dot{y}$ using the Lotka–Volterra equations (10):

$$\frac{df}{dt} = \left[kx\left(1 - \frac{y}{Y}\right)\right]\frac{h}{x}\left(1 - \frac{x}{X}\right) + \left[-hy\left(1 - \frac{x}{X}\right)\right]\frac{k}{y}\left(1 - \frac{y}{Y}\right)$$

$$= hk\left(1 - \frac{y}{Y}\right)\left(1 - \frac{x}{X}\right) - hk\left(1 - \frac{x}{X}\right)\left(1 - \frac{y}{Y}\right)$$

$$= 0.$$

So the function f is unchanging with time, which means that each path in
the phase plane must correspond to a different value of f. Thus the phase
paths are the contours of f traversed in some direction, as shown in
Figure 22. Hence the equilibrium point is a centre. The front cover of this
book shows the vector field of the Lotka–Volterra equations together with
coloured contours of f.

The above argument could use *any* function that is constant along paths in
order to show that the non-linear system has a centre. For a mechanical
system, an argument based on conservation of energy could produce a
function with the correct properties. For the Lotka–Volterra equations, the
function $f(x,y)$ arises from the following argument. Start with the
observation that

$$\frac{dy}{dx} = \frac{dy}{dt} \times \frac{dt}{dx} = \frac{\dot{y}}{\dot{x}}.$$

Now substitute for $\dot{x}$ and $\dot{y}$ using the Lotka–Volterra equations to get

$$\frac{dy}{dx} = \frac{-hy(1 - x/X)}{kx(1 - y/Y)}.$$

This first-order differential equation can be solved by separation of
variables:

$$k \int \frac{1 - y/Y}{y}\, dy = -h \int \frac{1 - x/X}{x}\, dx.$$

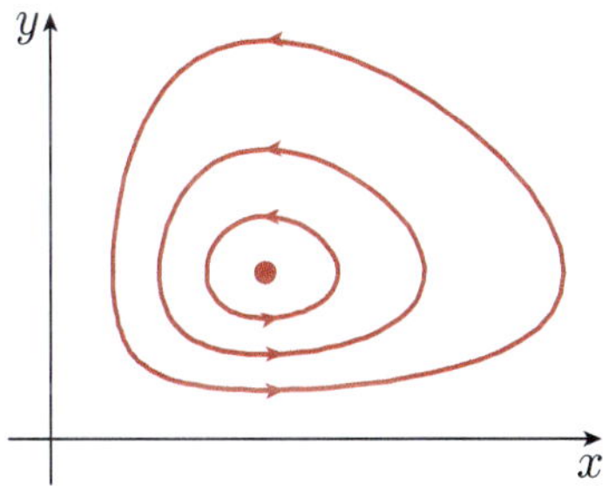

Figure 22   A phase portrait for the Lotka–Volterra equations; the dot marks an equilibrium point

Performing the integrations gives

$$k \ln y - \frac{k}{Y} y = -h \ln x + \frac{h}{X} x + C,$$

where C is a constant. Rearranging this equation gives $f(x, y)$.

4 Phase portraits

A system of non-linear differential equations may have several equilibrium points. Using the methods of the previous section, we can sketch paths in the neighbourhood of each equilibrium point. In this section we aim to knit together these sketches into a complete picture, or *portrait*, of the solutions of a system of differential equations.

We start by considering the equations of the predator–prey model again, and look at what general features about the phase plane can be deduced directly from the equations.

Exercise 24

Consider the system of differential equations defined in Exercise 4, namely

$$\dot{x} = x\left(1 - \frac{y}{2}\right), \quad \dot{y} = -\frac{1}{2}y\left(1 - \frac{x}{3}\right).$$

As we are interested in general features of these equations, consider both positive and negative values of x and y rather than restricting attention to those values that are physically reasonable for a population model (i.e. $x \geq 0$ and $y \geq 0$).

(a) For what values of x and y is $\dot{x}$ zero? In what regions of the plane is $\dot{x}$ positive? Sketch these regions on the phase plane.

(b) For what values of x and y is $\dot{y}$ zero? In which regions is $\dot{y}$ positive? Sketch these regions on the phase plane.

The regions of the plane that you sketched in Exercise 24 are significant. The regions where $\dot{x}$ is positive are the regions where the vector field arrows are pointing generally to the right. In these regions the phase paths will curve to the right. Similarly, the regions where $\dot{y}$ is positive are the regions where the phase paths curve upwards. This information is really useful when sketching phase portraits.

The boundaries of the regions are the key concept, as these boundaries are the only places where curves can change between curving leftwards/rightwards and upwards/downwards, so these are given a name.

The solutions of $\dot{x} = 0$ or $\dot{y} = 0$ are called **nullclines**. When we need to distinguish between the two sets of nullclines, the solutions of $\dot{x} = 0$ will be called the **nullclines for $\dot{x}$**, and the solutions of $\dot{y} = 0$ will be called the **nullclines for $\dot{y}$**.

Now we will use nullclines to plot the phase portrait of the predator–prey equations. The process that we go through will later be formalised in a procedure, and we will mark the steps of the procedure in the margin as we go along.

Example 7

Sketch the phase portrait of the system of differential equations

$$\dot{x} = x\left(1 - \frac{y}{2}\right), \quad \dot{y} = -\frac{1}{2}y\left(1 - \frac{x}{3}\right).$$

Solution

The first step in sketching a phase portrait is to find and classify the equilibrium points. In Example 1 we found that these equations have two equilibrium points, in this case (with $X = 3$ and $Y = 2$) $(0,0)$ and $(3,2)$. The given equations are the Lotka–Volterra equations for specific parameters; in Exercises 20 and 21 we classified the equilibrium points and found that $(0,0)$ is a saddle point and $(3,2)$ is a centre. These are marked in green in Figure 23 (see below).

◀ Equilibrium points ▶

The next step is to find the nullclines. In Exercise 24 we found that the nullclines for $\dot{x}$ are $x = 0$ and $y = 2$. These are marked by red lines in Figure 23. Also in Exercise 24 we found that the nullclines for $\dot{y}$ are $y = 0$ and $x = 3$. These are marked by blue lines in Figure 23.

◀ Nullclines ▶

Phase paths cross nullclines for $\dot{x}$ vertically (since $\dot{x} = 0$). We need to determine whether each crossing is upwards or downwards – this is determined by the sign of $\dot{y}$ on the nullcline. This sign can change only at the equilibrium points, as the right-hand sides of the system of differential equations are continuous. So all we need to do is determine the sign of $\dot{y}$ on either side of each equilibrium point. To do this, we compute the following values.

◀ Nullcline crossings ▶

(x,y)	$(0,1)$	$(0,-1)$	$(2,2)$	$(4,2)$
$\dot{y}$	$-1/2$	$1/2$	$-1/3$	$1/3$

In Figure 23, the positive signs are represented by upwards arrows, and the negative signs are represented by downward arrows.

Similarly, the phase paths cross the nullclines for $\dot{y}$ horizontally. We can evaluate $\dot{x}$ at points on either side of equilibrium points to find the direction of crossing as follows.

(x,y)	$(1,0)$	$(-1,0)$	$(3,1)$	$(3,3)$
$\dot{x}$	1	-1	$3/2$	$-3/2$

Again, we represent the signs of these values by arrows in Figure 23, where positive signs are represented by rightwards arrows and negative signs are represented by leftwards arrows.

◀ Nullcline regions ▶

The nullclines divide the phase plane into regions. In each region the vector field arrows will point in the same general direction (up/down and left/right) as $\dot{x}$ and $\dot{y}$ can change sign only on the nullclines. In Exercise 24 the regions shaded red are the regions where $\dot{x} > 0$, and the regions shaded blue are the regions where $\dot{y} > 0$. So the overlap regions are the regions where $\dot{x} > 0$ and $\dot{y} > 0$, that is, the arrows will point rightwards and upwards. Similarly, the regions that are not shaded by either colour are regions where $\dot{x} < 0$ and $\dot{y} < 0$, that is, regions where the arrows point leftwards and downwards. These signs are represented by arrows pointing north-east, north-west, south-east and south-west in Figure 23. Note that the arrows in this figure are purely representative of the general direction of the arrows in each region; they are not the computed arrows at this point.

We often abbreviate these directions to NE, NW, SE and SW.

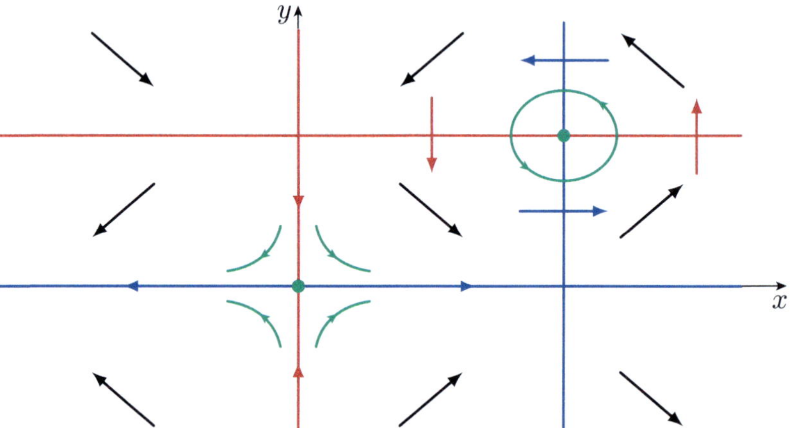

Figure 23   Accumulated information about the predator–prey equations. Nullclines for $\dot{x}$ are red, and nullclines for $\dot{y}$ are blue; arrows indicate crossings. Green dots and paths mark equilibrium points and their classification. Black arrows indicate the general direction in regions.

◀ Complete paths ▶

All that remains is to use this information to sketch typical paths in the phase plane. The aim here is not to fill the phase plane with paths, but to add sufficient paths to show the important features. This will usually involve showing the paths that start at equilibrium points (i.e. start infinitesimally close to but away from equilibrium points) and exploring each of the regions defined by the nullclines. The phase portrait for these equations is shown in Figure 24.

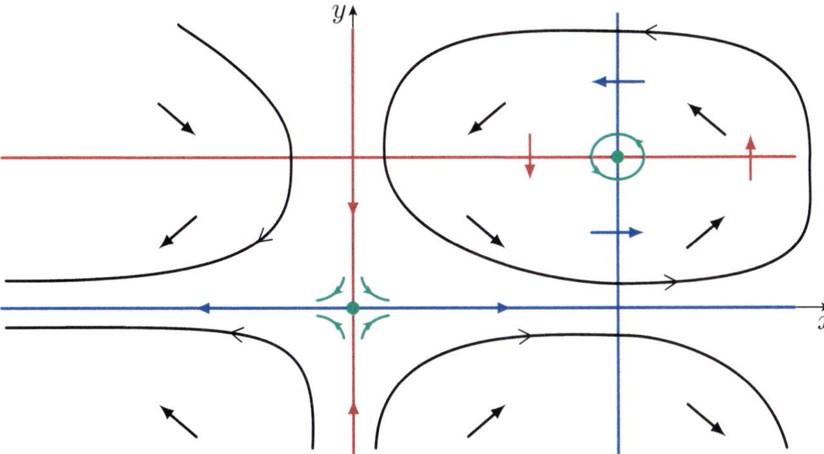

Figure 24 Nullclines for the predator–prey equations, together with the regions where $\dot{x} > 0$ and the regions where $\dot{y} > 0$

The steps used in Example 7 form the basis of a general procedure for sketching phase portraits.

Procedure 5   Sketching phase portraits

In order to sketch the phase portrait of a system of differential equations in x and y, carry out the following steps.

1. Use Procedure 4 to find and classify the equilibrium points. Mark these points on a sketch, and draw small sketches of the local behaviour of the paths near the equilibrium points.   ◀ Equilibrium points ▶

2. Find the nullclines by solving $\dot{x} = 0$ and $\dot{y} = 0$. Draw these on your sketch, using two different colours for the two sets of nullclines. (Note that the equilibrium points occur where nullclines of different colours intersect – this is a useful check.)   ◀ Nullclines ▶

3. For each equilibrium point on the nullclines for $\dot{x}$, evaluate the sign of $\dot{y}$ on either side. Mark this on your sketch by adding up or down arrows.   ◀ Nullcline crossings ▶

    For each equilibrium point on the nullclines for $\dot{y}$, evaluate the sign of $\dot{x}$ on either side. Mark this on your sketch by adding left or right arrows.

4. The nullclines divide the phase plane into several regions. Label each region with a NE, NW, SE or SW arrow to show the general direction of the arrows in that region.   ◀ Nullcline regions ▶

5. Extend the paths from the equilibrium points by curving the paths in the direction of the arrows in each region. Add any extra paths that do not start or end at equilibrium points, so that each region is crossed by at least one path.   ◀ Complete paths ▶

The following harder example illustrates the use of this procedure.

Example 8

Consider the system of differential equations that we looked at in Example 6, namely

$$\dot{x} = -4y + 2xy - 8, \quad \dot{y} = 4y^2 - x^2,$$

where we found that these equations have a source at $(4, 2)$ and a spiral sink at $(-2, -1)$.

Sketch the phase portrait.

Solution

The nullclines for $\dot{x}$ are given by

$$-4y + 2xy - 8 = 0.$$

Rearranging to make y the subject of this equation gives

$$y = \frac{4}{x - 2}.$$

So these nullclines are the two branches of a rectangular hyperbola obtained by taking the graph of $y = 1/x$, scaling by a factor 4, and translating 2 units to the right.

Now consider the nullclines for $\dot{y}$, which are given by

$$4y^2 - x^2 = 0.$$

This equation has solutions $x = 2y$ and $x = -2y$. So the nullclines in this case are a pair of straight lines.

A sketch of the nullclines is shown in Figure 25(a). One interesting feature to note from the diagram is that the equilibrium points (marked by dots) occur at the intersections of the red and blue lines. This will always happen, as the red lines correspond to $\dot{x} = 0$ and the blue lines correspond to $\dot{y} = 0$, so the intersection points are when both $\dot{x} = 0$ and $\dot{y} = 0$, that is, the equilibrium points.

Now we determine the directions of the nullcline crossings. We start with the nullclines for $\dot{x}$, which are the two branches of the rectangular hyperbola. Both equilibrium points lie on these nullclines, so we need to evaluate $\dot{y}$ at four points (one on each side of each equilibrium point). We choose the points to simplify the arithmetic.

(x, y)	$(-6, -1/2)$	$(0, -2)$	$(3, 4)$	$(6, 1)$
$\dot{y}$	-35	16	55	-32

The signs of these values are represented by red upwards and downwards pointing arrows in Figure 25(b).

Now consider the nullclines for $\dot{y}$. The nullcline $x = 2y$ also has two equilibrium points lying on it, but this time we can reduce the effort by considering only three points as the curve is continuous. So we calculate the following.

(x, y)	$(-4, -2)$	$(0, 0)$	$(6, 3)$
$\dot{x}$	16	-8	16

The signs of these values are represented by blue leftwards and rightwards pointing arrows in Figure 25(b).

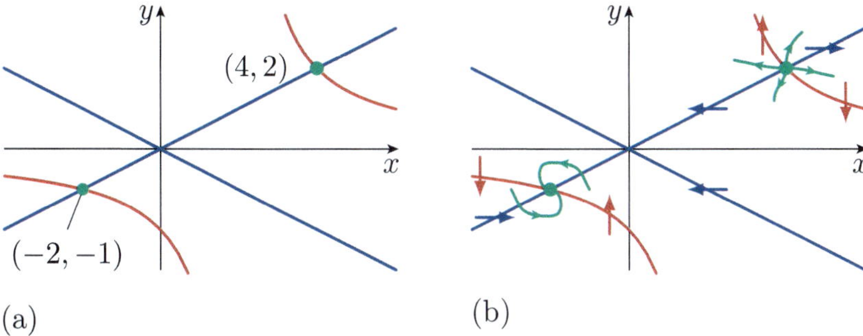

Figure 25   (a) Nullclines of the differential equations, with nullclines for $\dot{x}$ red and nullclines for $\dot{y}$ blue; the two dots mark the equilibrium points. (b) Nullclines plus the directions of crossings and paths in the neighbourhoods of equilibrium points.

The other nullcline for $\dot{y}$, namely the line $x = -2y$, has no equilibrium points on it. So the direction of crossing must be the same along the whole length of this line. So we need to evaluate $\dot{x}$ at only one point to determine this sign. In fact, we do not even have to do one evaluation, as this nullcline crosses the other nullcline for $\dot{y}$, and we know that the sign at the point of crossing is negative. So the sign of $\dot{y}$ must be negative along the whole length of this nullcline. This is marked in Figure 25(b).

We also know that the point $(4, 2)$ is a source and the point $(-2, -1)$ is a spiral sink. So the solution paths will flow outwards from the point $(4, 2)$ and spiral inwards to the point $(-2, -1)$. This information has also been added to Figure 25(b) as small sketches of the phase paths in the neighbourhoods of the equilibrium points. It helps to delay adding these sketches in the neighbourhoods of the equilibrium points until the nullcline crossings are known, as this helps to determine the sketch; for example, it indicates whether a spiral sink is a clockwise or anticlockwise spiral.

We now have a lot of information about the directions of the solution paths on the nullclines, and we can extend this to the regions of the plane with the nullclines as boundaries. Consider the region between the two branches of the red hyperbola. As the red hyperbola corresponds to the equation $\dot{x} = 0$, we know that $\dot{x}$ cannot change sign in this region, so all solution paths must progress leftwards. Similarly, we know that all solution paths in the top-right and bottom-left regions must progress to the right. Also, the plane is divided into four regions by the pair of blue nullclines. The solution paths must be generally pointing upwards in the north and south regions, and generally pointing downwards in the west and east regions. These are represented by arrows in the regions in Figure 26.

◀ Nullcline regions ▶

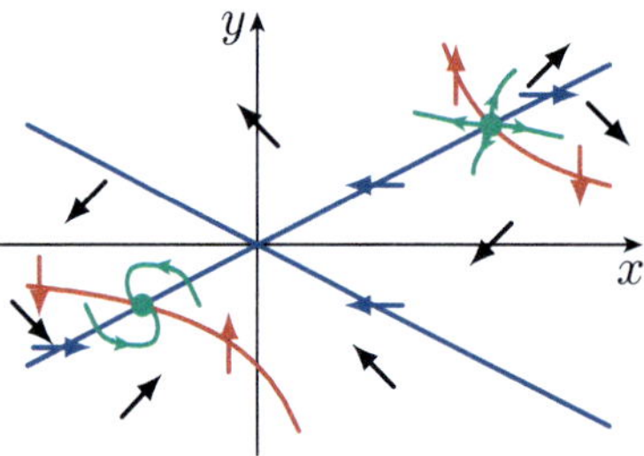

Figure 26 Collected information about the system of differential equations ready for sketching the phase portrait

◀ Complete paths ▶

Now that we have gathered all the information about the system of differential equations, we can sketch the phase portrait, as shown in Figure 27. You should not try to be too precise when drawing a sketch phase portrait. The aim is to convey the main features of the paths rather than precise details.

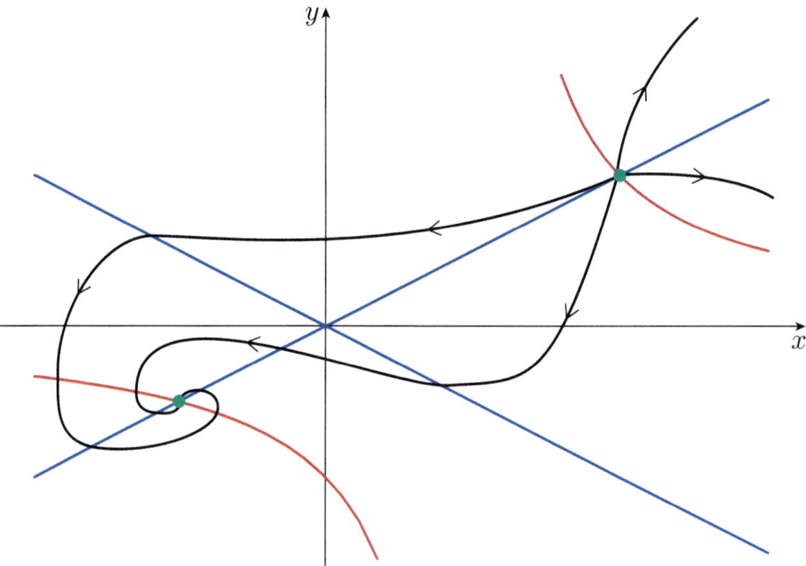

Figure 27 Phase portrait of the differential equations

The following exercise leads you through Procedure 5 step by step.

Exercise 25

Exercise 23 considered the system of differential equations

$$\dot{x} = (1 + x - 2y)x, \quad \dot{y} = (x - 1)y,$$

and found that this system has three equilibrium points, namely a saddle point at $(0,0)$, a sink at $(-1,0)$ and a spiral source at $(1,1)$.

◀ Equilibrium points ▶

(a) Mark the equilibrium points on a sketch, and draw short paths in the neighbourhood of each equilibrium point.

(b) Find the nullclines, and add these to your sketch.   ◀ Nullclines ▶

(c) Find the signs of the paths crossing the nullclines, and mark these on   ◀ Nullcline crossings ▶
your sketch.

(d) Add NW, NE, SW or SE arrows to each region of the phase plane   ◀ Nullcline regions ▶
separated by the nullclines on your sketch.

(e) Sketch the phase portrait of this system of equations.   ◀ Complete sketch ▶

This section concludes by looking ahead to phenomena that you may meet in your studies after completing this module. The first of these phenomena is when a solution tends to a repeating pattern rather than tending to an equilibrium point.

Recall that periodic solutions are represented by closed curves in the phase plane. Here we name a particular type of periodic solution.

A **limit cycle** of a system of differential equations is a closed solution curve to which nearby solution curves tend (either forwards or backwards in time).

An example will make this definition clearer, so consider the system of differential equations given by

$$\dot{x} = x(1 - x^2 - y^2) - y, \quad \dot{y} = y(1 - x^2 - y^2) + x. \tag{27}$$

The vector field plot for this system is shown in Figure 28.

This system of equations has a single equilibrium point at the origin, which is a spiral source. What is more interesting is what happens away from the origin. As the origin is a spiral source, the paths near the origin spiral outwards. In the figure, the arrows near the origin spiral anticlockwise and away from the origin.

However, near the edge of Figure 28 it can clearly be seen that the arrows correspond to paths that spiral inwards. This behaviour is because of the presence of the factor $(1 - x^2 - y^2)$ in both equations. For x and y small, this factor is positive, and this gives rise to paths spiralling outwards. For x and y large, the factor is negative, and this in turn gives rise to paths spiralling inwards. This factor is zero when $x^2 + y^2 = 1$, which corresponds to points on a circle of radius 1. For these points the system reduces to the simple linear system

$$\dot{x} = -y, \quad \dot{y} = x,$$

which has the solution $x(t) = \cos(t + \phi)$, $y(t) = \sin(t + \phi)$ that satisfies the initial condition $x = \cos\phi$, $y = \sin\phi$. So the circle $x^2 + y^2 = 1$ is a path in the phase plane. We have already seen that paths starting inside this circle spiral outwards towards the circle, and paths starting outside the circle spiral inwards towards the circle. Another way of saying the previous two sentences is to say that the circle $x^2 + y^2 = 1$ is a limit cycle.

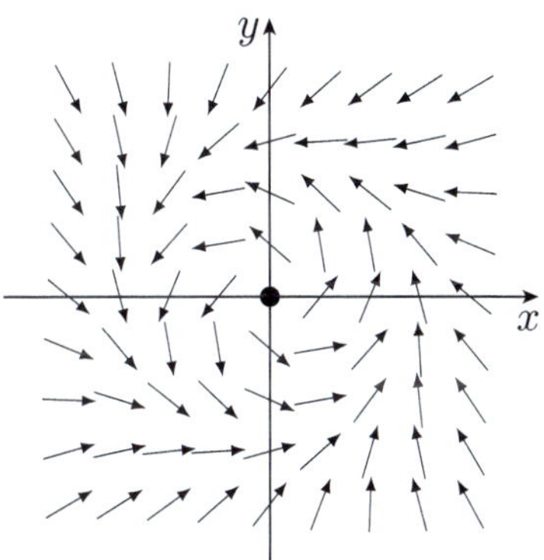

Figure 28   A vector field plot of a system of differential equations that contains a limit cycle

Another way of looking at equations (27) that is in some ways simpler is to consider transforming the equations into polar coordinates (r, θ) using the equations $x = r\cos\theta$ and $y = r\sin\theta$. In polar coordinates, equations (27) simplify to

$$\dot{r} = r(1 - r^2), \quad \dot{\theta} = 1.$$

Now the solution curves can be simply derived. The equation for θ can be integrated to give $\theta = t + C$ for some constant C, which means that the solution curves will spiral anticlockwise (the positive direction for polar coordinates) at a constant rate. The sign of $\dot{r}$ from the first equation shows that r is increasing for $r < 1$ and decreasing for $r > 1$. So the solutions of the r differential equation tend to $r = 1$ as t tends to infinity. This shows that the curve $r = 1$ (i.e. the circle with radius 1) is a limit cycle.

Methods of establishing the existence of limit cycles and determining their equations are beyond the scope of this module. However, when investigating systems of non-linear differential equations, you should be aware that limit cycles can exist.

For systems of two differential equations involving the time derivatives of two variables, the only features in phase portraits are equilibrium points and limit cycles (this celebrated result is known as the Poincaré–Bendixson theorem). In systems with three or more variables, other possibilities can occur, such as so-called chaotic motion. The study of chaos is a fascinating topic in any deeper study of systems of differential equations, but now seems to be a good point at which to finish this introduction to the topic.

Learning outcomes

After studying this unit, you should be able to:

- use a vector field to describe a pair of first-order non-linear differential equations, and use paths in the phase plane to represent the solutions

- understand how systems of differential equations arise from mathematical modelling, and in particular the modelling of populations of predators and prey

- convert a higher-order differential equation to a system of first-order differential equations

- find the equilibrium points of a system of non-linear differential equations

- find linear equations that approximate the behaviour of a system of non-linear differential equations near an equilibrium point, by calculating the Jacobian matrix

- determine whether an equilibrium point is stable or unstable

- use the eigenvalues and eigenvectors of the Jacobian matrix at an equilibrium point to classify an equilibrium point as a source, a sink, a star source, a star sink, an improper source, an improper sink, a spiral source, a spiral sink, a saddle, or a centre

- use nullclines to sketch the phase portrait of a system of differential equations.

Solutions to exercises

Solution to Exercise 1

The differential equation $\dot{x} = kx$ can be solved using the method of separation of variables to get, as $x > 0$,

$$\int \frac{dx}{x} = \int k\,dt.$$

Evaluating the integrals (and noting that x is positive) gives

$$\ln x = kt + A,$$

where A is a constant.

Using the given initial condition determines the value of the constant A:

$$\ln x_0 = k \times 0 + A.$$

Substituting for A and simplifying gives the required particular solution

$$\ln x = kt + \ln x_0.$$

We can write this as

$$\ln x - \ln x_0 = kt \quad \text{or} \quad \ln(x/x_0) = kt,$$

so

$$x = x_0 e^{kt}.$$

Solution to Exercise 2

The system of differential equations is

$$\dot{x} = x, \quad \dot{y} = -y.$$

These equations can be solved applying separation of variables to get the two equations

$$\int \frac{1}{x}\,dx = \int 1\,dt, \quad \int \frac{1}{y}\,dy = \int -1\,dt.$$

Performing the integrations and rearranging gives the general solution as

$$x(t) = Ce^{t}, \quad y(t) = De^{-t},$$

where C and D are constants.

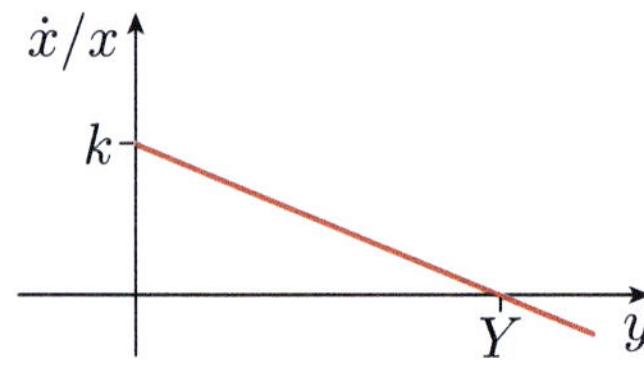

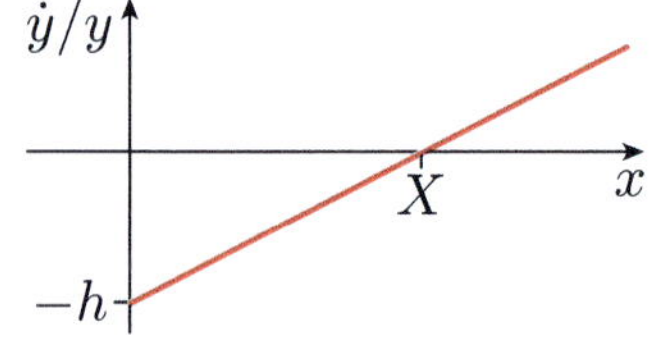

Solution to Exercise 3

Rearranging equation (8), we obtain

$$\frac{\dot{x}}{x} = k - \frac{k}{Y}\,y,$$

which is the equation of a straight line, as shown in the margin.

The proportionate growth rate $\dot{x}/x$ of rabbits decreases as the population y of foxes increases, becoming zero when $y = Y$. The population x of rabbits will increase if the population y of foxes is less than Y, but will decrease if y is greater than Y.

Similarly rearranging equation (9), we have

$$\frac{\dot{y}}{y} = -h + \frac{h}{X}\,x,$$

so the graph of $\dot{y}/y$ as a function of x is as shown in the margin.

The proportionate growth rate $\dot{y}/y$ of foxes increases linearly as the population x of rabbits increases. The population y of foxes will decrease if the population x of rabbits is less than X, but will increase if x is greater than X.

Solution to Exercise 4

(a) The Lotka–Volterra equations can be written as

$$\dot{\mathbf{x}} = \mathbf{u}(x, y),$$

where $\dot{\mathbf{x}} = (\dot{x} \quad \dot{y})^T$ and the vector field $\mathbf{u}(x, y)$ is given by

$$\mathbf{u}(x, y) = \begin{pmatrix} kx\left(1 - \dfrac{y}{Y}\right) \\ -hy\left(1 - \dfrac{x}{X}\right) \end{pmatrix}.$$

(b) The completed table is shown below.

x	y	$\mathbf{u}(x, y)$
0	0	$\mathbf{0}$
0	2	$(0 \quad -1)^T$
2	0	$(2 \quad 0)^T$
2	2	$(0 \quad -1/3)^T$
3	1	$(3/2 \quad 0)^T$
3	2	$\mathbf{0}$
3	3	$(-3/2 \quad 0)^T$
4	2	$(0 \quad 1/3)^T$

(c) The vectors are plotted in the following figure.

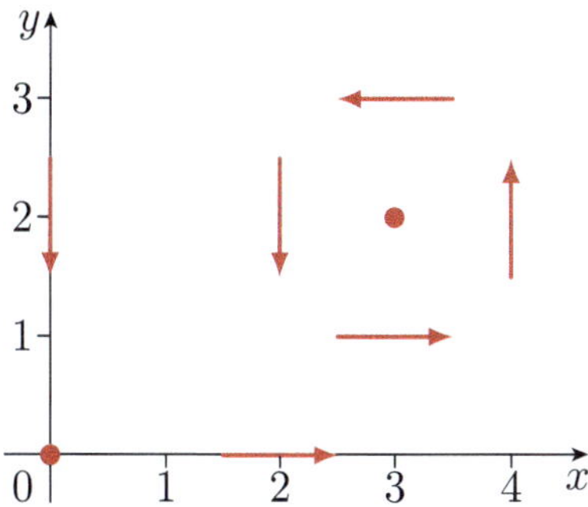

Solution to Exercise 5

(a) Let $y = \dot{x}$, so $\dot{y} = \ddot{x}$. Then the given equation becomes $\dot{y} + \sin x = 0$. This leads to the system of first-order equations

$$\dot{x} = y, \quad \dot{y} = -\sin x.$$

(b) As before, let $y = \dot{x}$, so $\dot{y} = \ddot{x}$. Then the given equation becomes $\dot{y} - 2y + x^2 = 0$, and the equivalent first-order system is

$$\dot{x} = y, \quad \dot{y} = 2y - x^2.$$

(c) Here we have different variables and a third-order equation, but the basic idea is still the same – that is, we introduce new variables for the derivatives of the independent variable. So let

$$v = \frac{du}{dx} \quad \text{and} \quad w = \frac{d^2u}{dx^2} = \frac{dv}{dx}.$$

In terms of these variables the given equation becomes

$$\frac{dw}{dx} = 6uv.$$

So the equivalent system of first-order equations is

$$\frac{du}{dx} = v, \quad \frac{dv}{dx} = w, \quad \frac{dw}{dx} = 6uv.$$

Solution to Exercise 6

Procedure 1 leads to the pair of simultaneous equations

$$x(20 - y) = 0,$$
$$y(10 - y)(10 - x) = 0.$$

From the first equation, either $x = 0$ or $y = 20$.

Substituting $x = 0$ into the second equation gives $y(10 - y) \times 10 = 0$, which gives $y = 0$ or $y = 10$. So $(0, 0)$ and $(0, 10)$ are equilibrium points.

Substituting $y = 20$ into the second equation gives $20 \times (-10) \times (10 - x) = 0$, which has solution $x = 10$. So $(10, 20)$ is an equilibrium point.

Hence the complete list of equilibrium points is $(0, 0)$, $(0, 10)$ and $(10, 20)$.

Solution to Exercise 7

Using Procedure 1, we must solve the pair of simultaneous equations

$$2x^2y + 7xy^2 + 2y + 1 = 0,$$
$$xy - x = 0.$$

In order to reduce the effort needed, we start by considering solutions of the second equation as it is much simpler. The second equation factorises as $x(y - 1) = 0$, which gives $x = 0$ or $y = 1$.

Substituting $x = 0$ into the first equation gives $2y + 1 = 0$, which has solution $y = -\frac{1}{2}$. So $(0, -\frac{1}{2})$ is an equilibrium point.

Substituting $y = 1$ into the first equation gives $2x^2 + 7x + 3 = 0$, which factorises as $(2x + 1)(x + 3) = 0$. So $x = -\frac{1}{2}$ or $x = -3$, which gives the equilibrium points $(-\frac{1}{2}, 1)$ and $(-3, 1)$.

So the complete list of equilibrium points is $(0, -\frac{1}{2})$, $(-\frac{1}{2}, 1)$ and $(-3, 1)$.

Solution to Exercise 8

The paths move towards the equilibrium point A, so it is stable.

The paths move away from the equilibrium point B, so it is unstable.

The paths move around the equilibrium point C thus stay in the neighbourhood of the point, so it is stable.

Solution to Exercise 9

We evaluate the various partial derivatives found in Example 3. At the equilibrium point $(0, 0)$, we obtain

$$u_x(0, 0) = k, \quad u_y(0, 0) = 0,$$
$$v_x(0, 0) = 0, \quad v_y(0, 0) = -h.$$

Thus the required linear approximation is

$$\begin{pmatrix} \dot{p} \\ \dot{q} \end{pmatrix} = \begin{pmatrix} k & 0 \\ 0 & -h \end{pmatrix} \begin{pmatrix} p \\ q \end{pmatrix},$$

giving the pair of equations

$$\dot{p} = kp, \quad \dot{q} = -hq.$$

Solution to Exercise 10

Here we have

$$u(x, y) = x(20 - y), \quad v(x, y) = y(10 - y)(10 - x),$$

giving partial derivatives

$$u_x(x, y) = 20 - y, \quad u_y(x, y) = -x,$$
$$v_x(x, y) = -y(10 - y), \quad v_y(x, y) = (10 - y)(10 - x) - y(10 - x).$$

So the Jacobian matrix of the vector field $\mathbf{u}(x, y)$ is

$$\mathbf{J}(x, y) = \begin{pmatrix} 20 - y & -x \\ -y(10 - y) & (10 - y)(10 - x) - y(10 - x) \end{pmatrix}.$$

At the equilibrium point $(10, 20)$ we have

$$\mathbf{J}(10, 20) = \begin{pmatrix} 0 & -10 \\ 200 & 0 \end{pmatrix},$$

so the linear approximation is

$$\begin{pmatrix} \dot{p} \\ \dot{q} \end{pmatrix} = \begin{pmatrix} 0 & -10 \\ 200 & 0 \end{pmatrix} \begin{pmatrix} p \\ q \end{pmatrix}.$$

(This is equivalent to the pair of equations

$$\dot{p} = -10q, \quad \dot{q} = 200p.)$$

Solution to Exercise 11

Solving the equations $3x + 2y - 8 = 0$ and $x + 4y - 6 = 0$, we obtain the equilibrium point $(2, 1)$.

Putting $x = 2 + p$ and $y = 1 + q$, we obtain the matrix equation

$$\begin{pmatrix} \dot{p} \\ \dot{q} \end{pmatrix} = \begin{pmatrix} 3 & 2 \\ 1 & 4 \end{pmatrix} \begin{pmatrix} p \\ q \end{pmatrix}.$$

(Note that this linear approximation near $(2, 1)$ is exact, since the original system is linear.)

Solution to Exercise 12

(a) To find the equilibrium points, we solve the simultaneous equations

$$0.5x - 0.000\,05x^2 = 0,$$
$$-0.1y + 0.0004xy - 0.01y^2 = 0.$$

Factorising these equations gives

$$0.5x(1 - 0.0001x) = 0,$$
$$-0.1y(1 - 0.004x + 0.1y) = 0.$$

The first equation gives

$$x = 0 \quad \text{or} \quad x = 10\,000.$$

If $x = 0$, the second equation is

$$-0.1y(1 + 0.1y) = 0,$$

which gives $y = 0$ or $y = -10$. As $y \geq 0$, only the first solution is possible. This leads to the equilibrium point $(0, 0)$.

If $x = 10\,000$, the second equation is

$$-0.1y(-39 + 0.1y) = 0,$$

which gives $y = 0$ or $y = 390$. So we have found two more equilibrium points, namely $(10\,000, 0)$ and $(10\,000, 390)$.

So this system has three equilibrium points, namely $(0, 0)$, $(10\,000, 0)$ and $(10\,000, 390)$.

(b) We have

$$u(x, y) = 0.5x - 0.000\,05x^2,$$
$$v(x, y) = -0.1y + 0.0004xy - 0.01y^2.$$

So the Jacobian matrix is

$$\mathbf{J}(x, y) = \begin{pmatrix} 0.5 - 0.0001x & 0 \\ 0.0004y & -0.1 + 0.0004x - 0.02y \end{pmatrix}.$$

(c) At the equilibrium point $(0, 0)$,

$$\mathbf{J}(0, 0) = \begin{pmatrix} 0.5 & 0 \\ 0 & -0.1 \end{pmatrix},$$

and the linearised approximations to the differential equations near this equilibrium point are

$$\begin{pmatrix} \dot{p} \\ \dot{q} \end{pmatrix} = \begin{pmatrix} 0.5 & 0 \\ 0 & -0.1 \end{pmatrix} \begin{pmatrix} p \\ q \end{pmatrix}.$$

At the equilibrium point $(10\,000, 0)$,

$$\mathbf{J}(10\,000, 0) = \begin{pmatrix} -0.5 & 0 \\ 0 & 3.9 \end{pmatrix},$$

and the linearised approximations to the differential equations near this equilibrium point are

$$\begin{pmatrix} \dot{p} \\ \dot{q} \end{pmatrix} = \begin{pmatrix} -0.5 & 0 \\ 0 & 3.9 \end{pmatrix} \begin{pmatrix} p \\ q \end{pmatrix}.$$

Finally, at the equilibrium point $(10\,000, 390)$,

$$\mathbf{J}(10\,000, 390) = \begin{pmatrix} -0.5 & 0 \\ 0.156 & -3.9 \end{pmatrix},$$

and the linearised approximations to the differential equations near this equilibrium point are

$$\begin{pmatrix} \dot{p} \\ \dot{q} \end{pmatrix} = \begin{pmatrix} -0.5 & 0 \\ 0.156 & -3.9 \end{pmatrix} \begin{pmatrix} p \\ q \end{pmatrix}.$$

Solution to Exercise 13

(a) As the given matrix is lower triangular, the eigenvalues can be read off the leading diagonal, so the eigenvalues are 1 and 3.

(b) As the eigenvalues are positive and distinct, the equilibrium point is a source.

Solution to Exercise 14

(a) The characteristic equation of the matrix of coefficients is

$$-\lambda(-3 - \lambda) + 2 = 0,$$

which simplifies to

$$\lambda^2 + 3\lambda + 2 = 0,$$

which factorises to give

$$(\lambda + 1)(\lambda + 2) = 0,$$

so the eigenvalues are $\lambda = -1$ and $\lambda = -2$.

(b) As the eigenvalues are negative and distinct, the equilibrium point is a sink.

Solution to Exercise 15

(a) The characteristic equation of the matrix of coefficients is

$$\begin{vmatrix} 1 - \lambda & 2 \\ 2 & -2 - \lambda \end{vmatrix} = 0,$$

or $\lambda^2 + \lambda - 6 = 0$, which factorises to give

$$(\lambda - 2)(\lambda + 3) = 0,$$

so the eigenvalues are $\lambda = 2$ and $\lambda = -3$.

The eigenvectors $(a \quad b)^T$ corresponding to $\lambda = 2$ satisfy the equations

$$-a + 2b = 0,$$
$$2a - 4b = 0.$$

So an eigenvector corresponding to the positive eigenvalue $\lambda = 2$ is $(2 \quad 1)^T$, and all the eigenvectors are along the line $q = \frac{1}{2}p$.

The eigenvectors $(a \quad b)^T$ corresponding to $\lambda = -3$ satisfy the equations

$$4a + 2b = 0,$$
$$2a + b = 0.$$

So an eigenvector corresponding to the negative eigenvalue $\lambda = -3$ is $(1 \quad -2)^T$, and all the eigenvectors are along the line $q = -2p$.

(b) The matrix of coefficients has a positive eigenvalue and a negative eigenvalue, so the equilibrium point is a saddle.

(c) There are two straight-line paths, namely $q = \frac{1}{2}p$ and $q = -2p$. On the line $q = \frac{1}{2}p$, the point $(p(t), q(t))$ moves away from the origin as t increases, because the corresponding eigenvalue is *positive*. On the line $q = -2p$, the point approaches the origin as t increases, because the corresponding eigenvalue is *negative*. This information, together with the knowledge that the equilibrium point is a saddle, allows us to sketch the phase portrait shown in the margin.

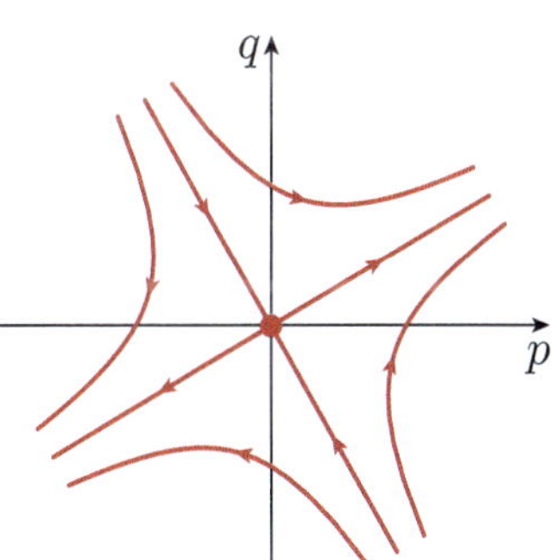

Solution to Exercise 16

(a) The characteristic equation of the matrix of coefficients is

$$(2 - \lambda)(-2 - \lambda) + 5 = 0,$$

that is, $\lambda^2 + 1 = 0$, so the eigenvalues are $\lambda = i$ and $\lambda = -i$.

(b) As the eigenvalues are imaginary, the equilibrium point is a centre.

Solution to Exercise 17

(a) The characteristic equation of the matrix of coefficients is

$$(1 - \lambda)^2 + 1 = 0,$$

that is, $\lambda^2 - 2\lambda + 2 = 0$, which has complex roots $\lambda = 1 + i$ and $\lambda = 1 - i$.

(b) As the eigenvalues are complex with positive real part, the equilibrium point is a spiral source.

Solution to Exercise 18

(a) As the matrix is diagonal, the eigenvalues can be read off the leading diagonal. So the eigenvalue is 2 (repeated).

In fact, the matrix is twice the identity matrix, so any vector transforms to twice itself. So any non-zero vector is an eigenvector. So we can choose $(1 \;\; 0)^T$ and $(0 \;\; 1)^T$ to be two linearly independent eigenvectors.

(b) The differential equations are

$$\dot{p} = 2p, \quad \dot{q} = 2q,$$

which have general solution

$$p(t) = Ce^{2t}, \quad q(t) = De^{2t},$$

where C and D are arbitrary constants.

(c) Eliminating t from the general solution, the equations of the paths are

$$q = \frac{D}{C}p = Kp \quad (C \neq 0),$$

where $K = D/C$ is also an arbitrary constant. So the paths are all straight lines passing through the origin.

The above analysis has neglected the possibility $C = 0$. In this case the path is $p = 0$, which is also a straight line passing through the origin, namely the q-axis.

(d) Both $p(t)$ and $q(t)$ are increasing functions of time, so the point $(p(t), q(t))$ moves away from the origin as t increases. So the equilibrium point is unstable.

Solution to Exercise 19

(a) The characteristic equation is

$$(2 - \lambda)^2 = 0,$$

so the matrix has the repeated eigenvalue $\lambda = 2$.

The eigenvectors $(a \;\; b)^T$ corresponding to this repeated eigenvalue satisfy the equations

$$0 = 0, \quad a = 0,$$

so all the eigenvectors take the form $(0 \quad k)^T$, where k is a (non-zero) constant. There is only one independent eigenvector, an obvious choice being $\mathbf{v} = (0 \quad 1)^T$.

(b) Using the solution to part (a) and Procedure 2 of Unit 6, we need to find the vector $\mathbf{b} = (c \quad d)^T$ that satisfies the equation

$$\begin{pmatrix} 0 & 0 \\ 1 & 0 \end{pmatrix} \begin{pmatrix} c \\ d \end{pmatrix} = \begin{pmatrix} 0 \\ 1 \end{pmatrix},$$

that is, $0 = 0$, $c = 1$.

So $\mathbf{b} = (1 \quad 0)^T$, and the general solution of the system of differential equations is

$$\begin{pmatrix} p \\ q \end{pmatrix} = C\left(\begin{pmatrix} 0 \\ 1 \end{pmatrix} t + \begin{pmatrix} 1 \\ 0 \end{pmatrix} \right) e^{2t} + D \begin{pmatrix} 0 \\ 1 \end{pmatrix} e^{2t},$$

that is,

$$p(t) = Ce^{2t}, \quad q(t) = Cte^{2t} + De^{2t},$$

where C and D are constants.

Solution to Exercise 20

The characteristic equation of the matrix of coefficients is

$$\lambda^2 + hk = 0.$$

Because h, k are positive, the eigenvalues are $\lambda = \pm i\sqrt{hk}$, so the equilibrium point is a centre.

Solution to Exercise 21

The eigenvalues of the matrix of coefficients are $\lambda = k$ and $\lambda = -h$, where h, k are positive. So the equilibrium point is a saddle. (In fact, in this case we have to restrict p and q to non-negative values, but this does not affect our conclusion.)

Solution to Exercise 22

(a) The characteristic equation is

$$\lambda^2 - 4\lambda + 13 = 0,$$

so the eigenvalues are $2 + 3i$ and $2 - 3i$, corresponding to the eigenvectors $(1 \quad -i)^T$ and $(1 \quad i)^T$, respectively.

As the eigenvalues are complex with a positive real component, the equilibrium point is a spiral source.

(b) As the equilibrium point of the linear approximation is not a centre, the corresponding equilibrium point of the non-linear system is also a spiral source.

Solution to Exercise 23

(a) The equilibrium points are given by

$$(1 + x - 2y)x = 0,$$
$$(x - 1)y = 0.$$

The second equation gives

$$x = 1 \quad \text{or} \quad y = 0.$$

When $x = 1$, substituting into the first equation gives

$$2 - 2y = 0,$$

which leads to $y = 1$. So $(1, 1)$ is an equilibrium point.

When $y = 0$, substituting into the first equation gives

$$(1 + x)x = 0,$$

hence $x = 0$ or $x = -1$. So we have found two further equilibrium points, namely $(0, 0)$ and $(-1, 0)$.

Thus we have three equilibrium points: $(1, 1)$, $(0, 0)$ and $(-1, 0)$.

(b) With the usual notation,

$$u(x, y) = (1 + x - 2y)x = x + x^2 - 2xy,$$
$$v(x, y) = (x - 1)y = xy - y.$$

So the Jacobian matrix is

$$\begin{pmatrix} u_x & u_y \\ v_x & v_y \end{pmatrix} = \begin{pmatrix} 1 + 2x - 2y & -2x \\ y & x - 1 \end{pmatrix}.$$

(c) At the point $(0, 0)$, the Jacobian matrix is

$$\begin{pmatrix} 1 & 0 \\ 0 & -1 \end{pmatrix},$$

so the linearised system is

$$\begin{pmatrix} \dot{p} \\ \dot{q} \end{pmatrix} = \begin{pmatrix} 1 & 0 \\ 0 & -1 \end{pmatrix} \begin{pmatrix} p \\ q \end{pmatrix}.$$

The eigenvalues of the matrix of coefficients are $\lambda = 1$ and $\lambda = -1$. As one of the eigenvalues is positive and the other is negative, the equilibrium point of the linearised system is a saddle.

At the point $(-1, 0)$, the Jacobian matrix is

$$\begin{pmatrix} -1 & 2 \\ 0 & -2 \end{pmatrix},$$

so the linearised system is

$$\begin{pmatrix} \dot{p} \\ \dot{q} \end{pmatrix} = \begin{pmatrix} -1 & 2 \\ 0 & -2 \end{pmatrix} \begin{pmatrix} p \\ q \end{pmatrix}.$$

The eigenvalues of the matrix of coefficients are $\lambda = -1$ and $\lambda = -2$. As the eigenvalues are negative and distinct, the equilibrium point of the linearised system is a sink.

At the point $(1, 1)$, the Jacobian matrix is

$$\begin{pmatrix} 1 & -2 \\ 1 & 0 \end{pmatrix},$$

so the linearised system is

$$\begin{pmatrix} \dot{p} \\ \dot{q} \end{pmatrix} = \begin{pmatrix} 1 & -2 \\ 1 & 0 \end{pmatrix} \begin{pmatrix} p \\ q \end{pmatrix}.$$

The characteristic equation of the matrix of coefficients is

$$\lambda^2 - \lambda + 2 = 0.$$

The roots of this quadratic equation are

$$\lambda = \tfrac{1}{2}(1 \pm i\sqrt{7}),$$

so the eigenvalues are complex with a positive real component.

Hence the equilibrium point of the linearised system is a spiral source.

(d) As none of the equilibrium points of the linearised systems found in part (c) are centres, the behaviour of the original non-linear system near the equilibrium points is the same as that of the linear approximations. In other words,

$$\begin{aligned}
(0, 0) \quad &\text{is a saddle,} \\
(-1, 0) \quad &\text{is a sink,} \\
(1, 1) \quad &\text{is a spiral source.}
\end{aligned}$$

Solution to Exercise 24

(a) To answer this part of the exercise we use the equation

$$\dot{x} = x\left(1 - \frac{y}{2}\right).$$

As this equation is already factorised, we deduce that $\dot{x} = 0$ when $x = 0$ or when $y = 2$.

For $\dot{x}$ to be positive, either both terms in the product must be positive or both terms must be negative. Both terms are positive when $x > 0$ and $y < 2$. Both terms are negative when $x < 0$ and $y > 2$. This gives two regions where the growth rate $\dot{x}$ is positive.

The figure in the margin is a sketch of these lines and regions. The two lines where $\dot{x}$ is zero are marked, and the two regions where $\dot{x}$ is positive are shaded.

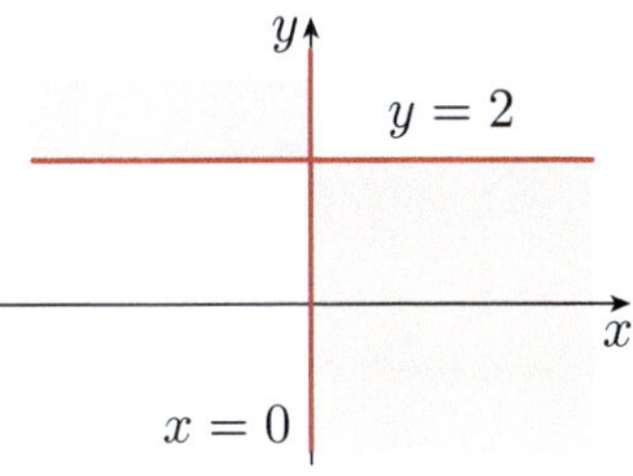

(b) To answer this part of the exercise we use the equation

$$\dot{y} = -\frac{1}{2}y\left(1 - \frac{x}{3}\right).$$

Again, this equation is already factorised, so $\dot{y} = 0$ has solutions $y = 0$ or $x = 3$.

For $\dot{y}$ to be positive, the terms in the product $y(1 - x/3)$ must have opposite signs. This occurs when $y > 0$ and $x > 3$ and also when $y < 0$ and $x < 3$.

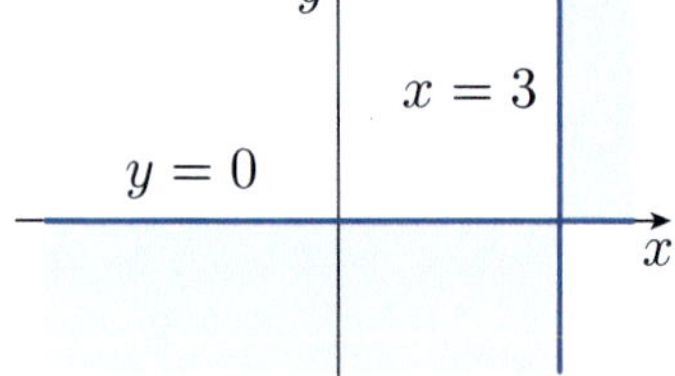

The figure in the margin is a sketch of these lines and regions. The two lines where $\dot{y}$ is zero are marked, and the two regions where $\dot{y}$ is positive are shaded.

Solution to Exercise 25

(a) The equilibrium points are marked in the figure in the margin.

As described in Example 8, it is easiest to defer marking the paths near equilibrium until after the nullcline crossings have been determined. In this case the nullcline crossings indicate whether the spiral source at $(1, 1)$ is a clockwise or anticlockwise spiral.

(b) The nullclines for $\dot{x}$ are $x = 0$ and $1 + x - 2y = 0$ (which is the line $y = (1 + x)/2$). These are shown as the two red lines in the figure in the margin.

The nullclines for $\dot{y}$ are $y = 0$ and $x = 1$, which are shown as the two blue lines in the figure in the margin.

(c) The nullcline $x = 0$ has one equilibrium point on it, so to determine the direction of the crossings, we evaluate $\dot{y}$ at two points.

(x, y)	$(0, 1)$	$(0, -1)$
$\dot{y}$	-1	1

The signs of these values are marked by the red arrows along the y-axis in the figure in the margin.

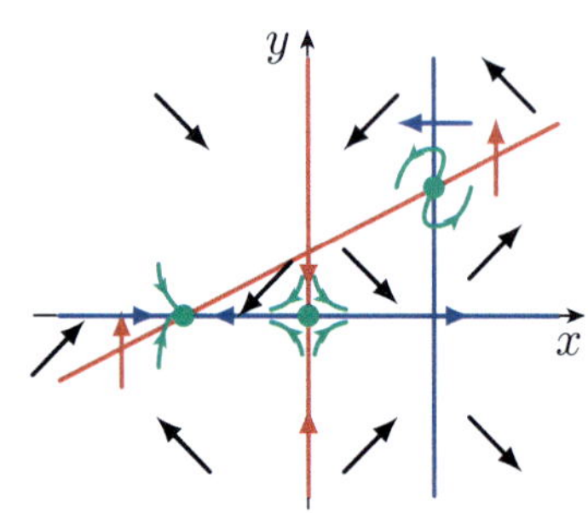

The nullcline $y = (1 + x)/2$ has two equilibrium points on it, so to determine the direction of the crossings, we evaluate $\dot{y}$ at three points.

(x, y)	$(-2, -1/2)$	$(0, 1/2)$	$(3, 2)$
$\dot{y}$	$3/2$	$-1/2$	4

These are represented by up and down arrows marked in the figure in the margin.

The nullcline $y = 0$ has two equilibrium points on it, so we need to evaluate $\dot{x}$ at three points.

(x, y)	$(-2, 0)$	$(-1/2, 0)$	$(1, 0)$
$\dot{x}$	2	$-1/4$	2

These are represented by left and right arrows marked in the figure in the margin.

The nullcline $x = 1$ has one equilibrium point, so we need to evaluate $\dot{x}$ at two points.

(x, y)	$(1, 0)$	$(1, 2)$
$\dot{x}$	2	-2

These are represented by left and right arrows marked in the figure in the margin.

(d) Arrows indicating the general directions in which the phase paths curve in each region are shown in the diagram in the margin.

(e) The phase portrait for these equations is shown below.

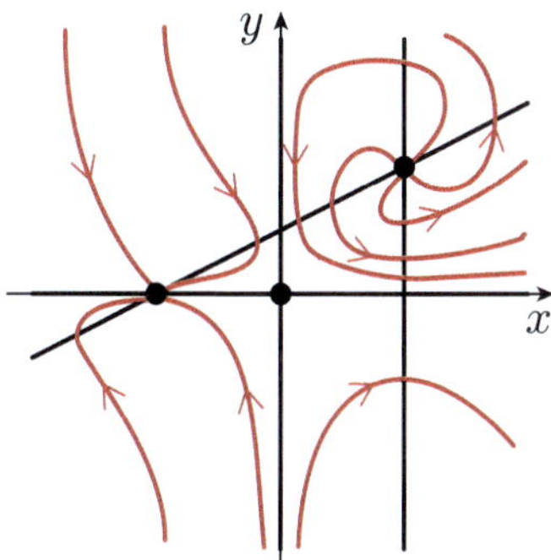

Fourier series

Introduction

In Unit 7 you saw that many functions can be approximated by a Taylor
polynomial

$$f(x) \simeq f(x_0) + f'(x_0)\,(x - x_0) + \frac{1}{2!}f''(x_0)\,(x - x_0)^2 + \cdots$$

$$+ \frac{1}{n!}f^{(n)}(x_0)\,(x - x_0)^n.$$

It is often the case that a small number of terms gives a useful
approximation, and it is tempting to ask whether the approximation may
be made exact by taking an *infinite* number of terms – in other words, is

$$f(x) = \sum_{n=0}^{\infty} \frac{1}{n!}f^{(n)}(x_0)\,(x - x_0)^n$$

true? This is indeed true for sufficiently smooth functions, but it is not
necessarily true for all functions. An example for which the above formula
is true is when $f(x) = \exp(x)$ and $x_0 = 0$. In this case we have
$f^{(n)}(x_0) = e^0 = 1$, so

$$e^x = \sum_{n=0}^{\infty} \frac{1}{n!}\,x^n,$$

for any real number x.

In this unit we are primarily concerned not with polynomial functions
(which, if not constant, become numerically very large as $x \to \pm\infty$), but
with *periodic* functions, such as $\sin x$ and $\cos x$. A great deal of beautiful
mathematics has arisen from the analysis that we will describe, and there
are also important practical benefits.

One example arises directly from Unit 10, where you studied differential
equations modelling forced and damped oscillations. You saw how they
could be used to predict the response of various mechanical and electrical
systems. In particular, you saw how they responded to a sinusoidal forcing
term like $\cos t$, with graph as in Figure 1.

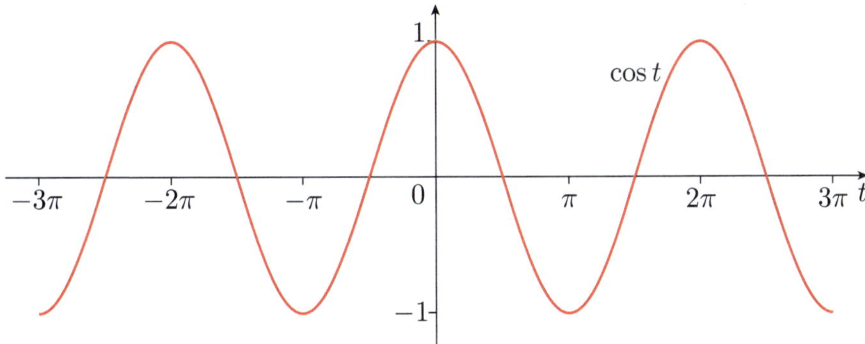

Figure 1   The sinusoidal function $\cos t$

Such forcing terms occur frequently in applications; for instance, they model the force acting on the suspensions of cars travelling on bumpy roads and the effect of radio signals acting on electrical circuits. The resulting model leads to a differential equation of the form

$$m\ddot{x} + r\dot{x} + kx = P\cos(\Omega t),$$

with steady-state solution

$$x(t) = PM\cos(\Omega t + \phi).$$

The amplitude magnification M and the phase angle ϕ are rather complicated functions of the forcing frequency Ω, which we need not consider here.

However, a forcing term, though periodic, may be more complicated than a purely sinusoidal function. Figures 2, 3 and 4 depict periodic functions $b(t)$, $g(t)$ and $h(t)$ that are reasonably easy to visualise and describe.

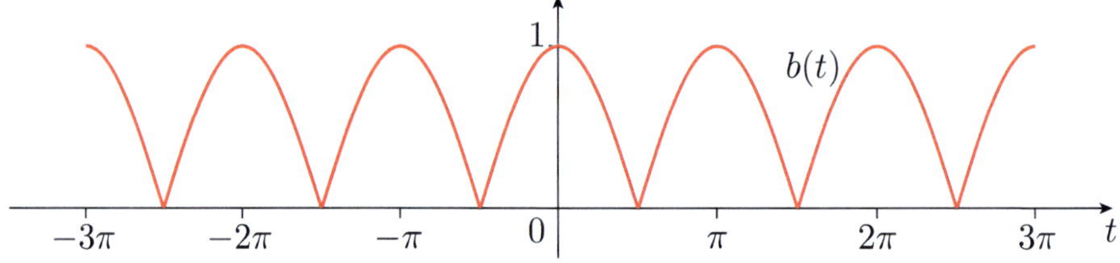

Figure 2   The periodic function $b(t)$, which resembles a series of bumps

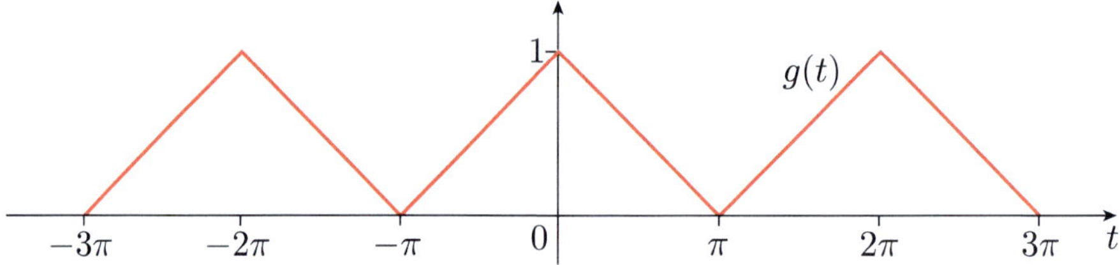

Figure 3   The periodic function $g(t)$, also known as a *sawtooth function*

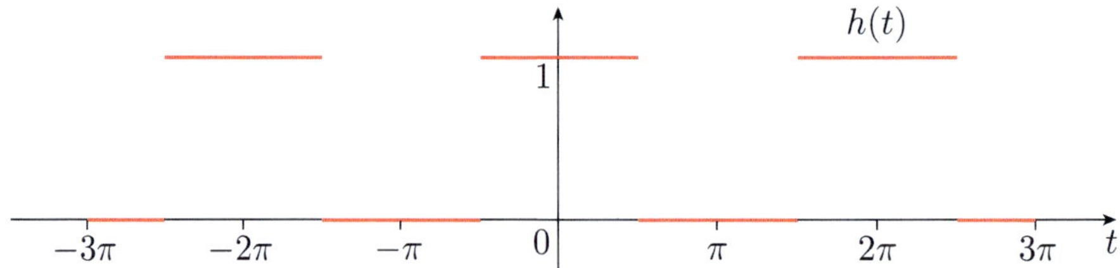

Figure 4   The periodic function $h(t)$, also known as a *square-wave function*

The graph in Figure 2 is similar to that of $\cos t$, except that all the values are positive. The function $b(t)$ is described by

$$b(t) = |\cos t|.$$

This graph differs from that of the cosine function, which turns smoothly. Here the direction changes abruptly every time the graph reaches the t-axis. However, $b(t)$ is still continuous in that there are no sudden jumps in the function value as t increases smoothly; it is possible to draw the graph without taking the pen off the paper.

The graph of $g(t)$ in Figure 3 is also continuous, but with abrupt changes of direction at both the highest and lowest values. Between the points where the direction changes, the graph is a straight line. The whole graph looks rather like the blade of a saw, so $g(t)$ is known as a **sawtooth function**. You will meet this function again in Unit 14, where it will be used to model the initial displacement of a plucked guitar string.

The graph in Figure 4 shows a function that takes the value 1 whenever t lies in the interval $[(2k - \frac{1}{2})\pi, (2k + \frac{1}{2})\pi]$ for some integer k, and the value 0 otherwise, that is,

$$h(t) = \begin{cases} 1 & \text{for } (2k - \frac{1}{2})\pi \leq t \leq (2k + \frac{1}{2})\pi \quad (k \in \mathbb{Z}), \\ 0 & \text{otherwise.} \end{cases}$$

This graph differs quite radically from those of $\cos t$, $b(t)$ and $g(t)$. Here there are abrupt jumps in the function value itself (rather than merely abrupt changes of direction) at the points $t = (k + \frac{1}{2})\pi$. The function $h(t)$ is discontinuous, while $\cos t$, $b(t)$ and $g(t)$ are continuous. The function $h(t)$ is known as a **square-wave function**.

At present we have no way of predicting the response of a mechanical or electrical system to a forcing function like $b(t)$, $g(t)$ or $h(t)$ (although such systems are extremely common). Fourier series provide the answer. The differential equation

$$m\ddot{x} + r\dot{x} + kx = P\cos(\Omega t)$$

is *linear*. Therefore the principle of superposition tells us that if   See Unit 1.

$$x = P_1 M_1 \cos(\Omega_1 t + \phi_1)$$

is a solution of

$$m\ddot{x} + r\dot{x} + kx = P_1 \cos(\Omega_1 t),$$

and

$$x = P_2 M_2 \cos(\Omega_2 t + \phi_2)$$

is a solution of

$$m\ddot{x} + r\dot{x} + kx = P_2 \cos(\Omega_2 t),$$

then

$$x = P_1 M_1 \cos(\Omega_1 t + \phi_1) + P_2 M_2 \cos(\Omega_2 t + \phi_2)$$

is a solution of

$$m\ddot{x} + r\dot{x} + kx = P_1 \cos(\Omega_1 t) + P_2 \cos(\Omega_2 t).$$

If we could express $b(t)$ as a linear combination of cosine functions, then we would be able to apply this idea to find a solution of

$$m\ddot{x} + r\dot{x} + kx = b(t)$$

(and similarly for $g(t)$ and $h(t)$). This is precisely what we will do in this unit, except that the linear combinations will involve an *infinite* number of terms.

As you will see, the functions $b(t)$, $g(t)$ and $h(t)$ introduced above correspond, respectively, to the infinite sums

$$B(t) = \tfrac{2}{\pi}\left(1 + \tfrac{2}{1\times 3}\cos 2t - \tfrac{2}{3\times 5}\cos 4t + \tfrac{2}{5\times 7}\cos 6t - \dots\right), \qquad (1)$$

$$G(t) = \tfrac{1}{2} + \tfrac{4}{\pi^2}\cos t + \tfrac{4}{9\pi^2}\cos 3t + \tfrac{4}{25\pi^2}\cos 5t + \tfrac{4}{49\pi^2}\cos 7t + \cdots, \qquad (2)$$

$$H(t) = \tfrac{1}{2} + \tfrac{2}{\pi}\cos t - \tfrac{2}{3\pi}\cos 3t + \tfrac{2}{5\pi}\cos 5t - \tfrac{2}{7\pi}\cos 7t + \cdots. \qquad (3)$$

We will use the convention that a function named with an upper-case letter (such as $B(t)$) corresponds to a given function with a lower-case letter (such as $b(t)$). Later in the unit the precise nature of this correspondence will be stated, and it will turn out that for almost all values of t these two functions will have equal values.

Infinite sums like $B(t)$, $G(t)$ and $H(t)$ are called **Fourier series** (named after Joseph Fourier; see Figure 5). Successive terms in each sum are functions that belong to a family of sinusoidal functions whose frequencies are related. The sums are infinite in the sense that they do not stop after a finite number of terms, though in practice we take only as many terms as are needed to make the result as accurate as we require.

Fourier series are not just of interest in the analysis and application of damped forced oscillations, but are widely applicable and are of fundamental theoretical importance. Whenever a system exhibits variation at a range of frequencies, it is sensible to see if this variation can be explained by some combination of sinusoidal terms.

Figure 5   The French mathematician, physicist and historian Joseph Fourier (1768–1830)

In Unit 14 you will look at transverse vibrations of guitar strings and at the conduction of heat along metal rods. These effects will be modelled by differential equations involving the partial derivatives that you met in Unit 7. In the case of the vibrating systems, there are sinusoidal solutions with a range of frequencies corresponding to the normal modes of Unit 11. These can be combined to find particular solutions. However, it is not so obvious that solutions to the heat-conduction problem can also be found as sums of sinusoidal terms. This was one of Fourier's many great discoveries.

Fourier realised that by considering series of sinusoidal functions, he could approximate most periodic functions. It is these series that we introduce and explore in this unit. You have already studied the mathematics that you need. Here all we have to do is draw it together to obtain powerful results.

In Section 1 we introduce Fourier series. This first involves a discussion of families of periodic functions and their periods, frequencies and fundamental intervals, and of even and odd functions. Section 2 is devoted to the task of calculating the Fourier series for a particular function, and in doing so we establish the principles for finding *any* Fourier series. Section 3 extends the results in Section 2 to find general formulas for the Fourier series for both even and odd functions with any given period. In Section 4 we extend these formulas to deal with functions that are neither even nor odd. Finally, we look at the problem of extending a function defined on an interval so that the extended function is periodic and has desirable properties. It is this final technique that will be used in the next unit to solve partial differential equations.

1 Introducing Fourier series

In this section we ask what kinds of functions can be expressed as Fourier series. You will see that a requirement is that the function is periodic. This section investigates a family of periodic functions, then looks at general even and odd functions. For a periodic function, we introduce the notions of *period, angular frequency* and *fundamental interval*.

1.1 Families of cosine functions

Suppose that we *define* the function $G(t)$ by the following Fourier series (an infinite series of cosine terms):

$$G(t) = \frac{1}{2} + \frac{4}{\pi^2} \cos t + \frac{4}{9\pi^2} \cos 3t + \frac{4}{25\pi^2} \cos 5t + \frac{4}{49\pi^2} \cos 7t + \cdots . \quad (4)$$

We now investigate some properties of $G(t)$, *without* assuming any connection with the sawtooth function $g(t)$.

The individual cosine terms in the sum are periodic, so it is not surprising to find that the sum $G(t)$ is also periodic.

Exercise 1

Let $G(t)$ be defined as in equation (4). Find $G(t + 2\pi)$ in terms of $G(t)$.

You saw in Exercise 1 that the function $G(t)$ defined by equation (4) is periodic with period 2π, just like the cosine function $\cos t$. But what of the individual terms in the sum? Apart from the constant term, they are multiples of

$$\cos t, \quad \cos 3t, \quad \cos 5t, \quad \cos 7t, \quad \ldots \qquad (5)$$

Exercise 2

What are the angular frequencies and periods of the functions in sequence (5)?

(Recall that in the expression $\cos(\omega t)$, the constant ω is called the **angular frequency**. The angular frequency is related to the period using the fact that the cosine function will repeat when its argument increases by 2π, so $\omega\tau = 2\pi$, where τ is the period, that is, $\tau = 2\pi/\omega$.)

From Exercise 2 you can see that we have a family of cosine functions where all the angular frequencies are integer multiples of the smallest angular frequency 1, and whose periods are integer fractions of the *fundamental period* 2π. We have seen that this is the period of the function $G(t)$, since all the component functions will have repeated after this time – some having repeated several times.

More generally (as you may recall from Unit 9), any function $f(t)$ is said to be **periodic** if it repeats regularly, that is, if there is some positive value λ such that for all t, $f(t + \lambda) = f(t)$. In this case, it is also true that for all t, $f(t + 2\lambda) = f(t + \lambda) = f(t)$, so 2λ could be taken as a period for $f(t)$ instead of λ, and in general, $n\lambda$ could be taken as the period (where n is any positive integer). The **fundamental period** of a periodic function is the *smallest* possible (positive) value for the period.

In applications, the fundamental period is far more important than the other periods. For this reason, 'fundamental period' is usually shortened to simply *period*. For example, the fundamental period of a pendulum (the time that it takes to swing to and fro) is usually called *the* period of the pendulum. We will occasionally use this shorthand when there is no risk of confusion. If we talk about *the period* of a function, then we mean its fundamental period.

In this unit, time is the independent variable of functions, and τ is used to denote the period of a periodic function.

Example 1

Let $f(t) = \cos 4t + 3\cos 6t$. What are the angular frequencies and corresponding periods of the component functions? What is the period of the function $f(t)$?

Solution

The angular frequencies of the component functions are 4 and 6. Their corresponding periods are $\frac{\pi}{2}$ and $\frac{\pi}{3}$, respectively. Hence the period of the combined function $f(t)$ is $\tau = \pi$ (as this is the shortest time that is an integer multiple of both $\frac{\pi}{2}$ and $\frac{\pi}{3}$). After this time, the first cosine term will have completed two cycles, while the other cosine term will have completed three.

Example 1 suggests that the period of a sum of sinusoidal terms is the least integer common multiple of the periods of the component functions. This period gives the first time after which *all* the component functions have repeated.

Exercise 3

Let $f(t) = 2\cos \pi t + 3\cos \frac{3\pi}{2} t - \cos 2\pi t$. What are the angular frequencies and corresponding periods of the component functions? What is the period of the function $f(t)$?

A Fourier series is an infinite sum of sinusoidal terms each of which is periodic, so, as has been exemplified above, a Fourier series is also periodic. Hence a sensible restriction on a function that is to be described by a Fourier series is that it should itself be periodic. However, as you have seen above, if you are interested in obtaining Fourier series for a function with period τ, then you must consider not only sinusoidal functions with period τ in the infinite sum, but also sinusoidal functions with fractional periods

$$\frac{\tau}{2}, \ \frac{\tau}{3}, \ \frac{\tau}{4}, \ \frac{\tau}{5}, \ \ldots,$$

since functions with these periods also repeat after time τ. Corresponding to the periods $\tau, \frac{\tau}{2}, \frac{\tau}{3}, \frac{\tau}{4}, \frac{\tau}{5}, \ldots$ are the angular frequencies

$$\frac{2\pi}{\tau}, \ \frac{4\pi}{\tau}, \ \frac{6\pi}{\tau}, \ \frac{8\pi}{\tau}, \ \frac{10\pi}{\tau}, \ \ldots.$$

So, for example, for a Fourier series of cosine functions, you must consider the family of functions

$$C_n(t) = \cos\left(\frac{2n\pi t}{\tau}\right), \qquad \text{where } n \text{ is a positive integer.} \tag{6}$$

Since these functions repeat after a time τ, we do not need to draw their graphs for all values of t. We can restrict our attention to any interval of length τ. We will choose the interval $\left[-\frac{\tau}{2}, \frac{\tau}{2}\right]$ as it has the correct length and is centred on the origin. In general, any interval whose length is the fundamental period can be chosen as the **fundamental interval** for functions of that period. The graph of the function $C_1(t)$ is shown in Figure 6.

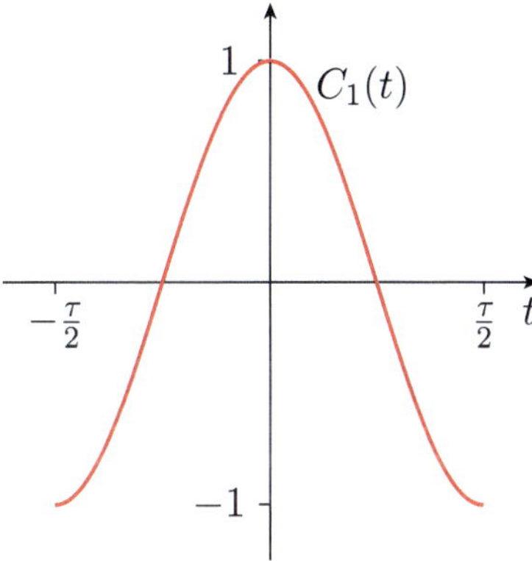

Figure 6   The function $C_1(t)$ plotted over its fundamental interval

---

Exercise 4

Sketch the graphs of the functions $C_2(t)$ and $C_3(t)$ on the fundamental interval $\left[-\frac{\tau}{2}, \frac{\tau}{2}\right]$.

What happens if you try to define $C_0(t)$ using formula (6)?

We have now obtained a family of cosine functions that repeat after a time τ, including the constant function $C_0(t) = 1$. (Whatever period τ is chosen, it is trivially true that a constant function has the same value after that period.) You will see in Section 2 how this family can be used to obtain Fourier series. But first, in Subsection 1.2, we need to digress slightly to discuss even and odd functions as these ideas are used to simplify later calculations.

1.2   Even and odd functions

The previous subsection dealt solely with cosine functions, but sine functions are also periodic, so why have we not used them? In fact, there is a distinguishing feature that separates these two families of functions. We will investigate this difference as it leads to a simplification of calculations later in the unit.

Figure 7 shows the graphs of a cosine function and a sine function, namely

$$C_1(t) = \cos\left(\frac{2\pi t}{\tau}\right), \quad S_1(t) = \sin\left(\frac{2\pi t}{\tau}\right).$$

They both have period τ and hence the same fundamental interval.

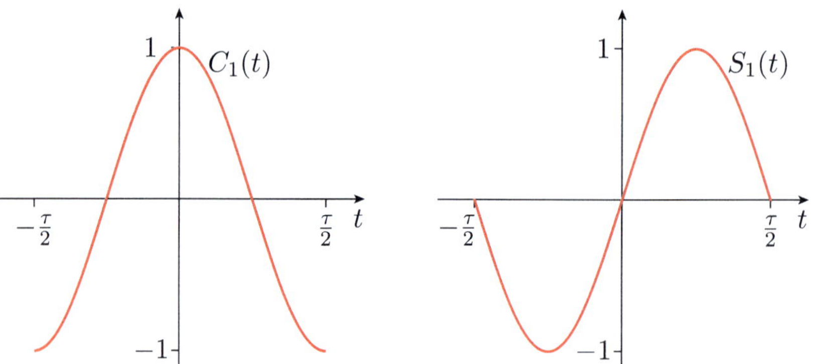

Figure 7   The functions $C_1(t)$ and $S_1(t)$ compared on their fundamental interval

Both graphs exhibit symmetry. You can see that the graph of the cosine function $C_1(t)$ takes the same values at corresponding points on either side of the vertical axis. We say that the function is *even*. By contrast, in the graph of the sine function $S_1(t)$, the values at corresponding points on either side of the vertical axis have the same magnitude but opposite signs. We say that the function is *odd*.

A function need not be either even or odd, as you will see below.

> The function $f(t)$ is an **even function** if
>
> $$f(-t) = f(t) \quad \text{for all values of } t;$$
>
> it is an **odd function** if
>
> $$f(-t) = -f(t) \quad \text{for all values of } t.$$

Example 2

Suppose that the function $f(t)$ is defined by $f(t) = t^2$. Is this function even, odd, or neither even nor odd?

Solution

Since $f(-t) = (-t)^2 = t^2 = f(t)$ for all t, the function is even.

Exercise 5

Suppose that the function $g(t)$ is defined by $g(t) = t^3$. Is this function even, odd, or neither even nor odd?

Exercise 6

If $f(t)$ and $g(t)$ are both odd functions, show that the function $k(t)$ defined by

$$k(t) = f(t) + g(t)$$

is also an odd function.

The graphs of the functions $f(t)$ and $g(t)$ defined in Example 2 and Exercise 5, and shown in Figure 8, should make the definitions clearer.

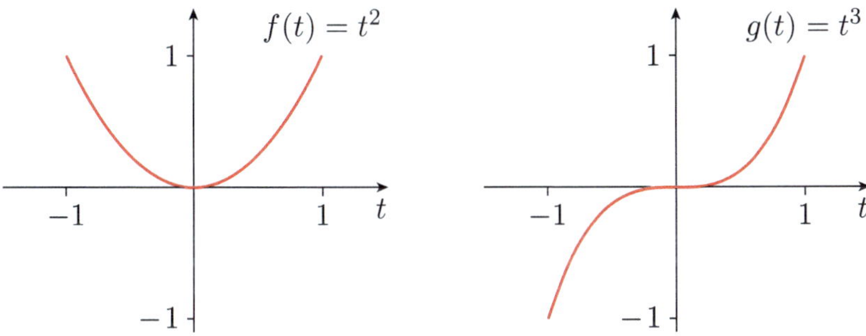

Figure 8   Graphs of the even function $f(t) = t^2$ and the odd function $g(t) = t^3$

For the even function $f(t)$, the same values appear on either side of the vertical axis, so the graph has reflection symmetry about this line. For the odd function $g(t)$, the values on either side of the vertical axis have opposite signs, so the graph has rotational symmetry through the angle π about the origin.

Generalising from the function $f(t) = t^2$, we can state that polynomial functions where *all* the powers are even are themselves even functions. Similarly, polynomial functions where *all* the powers are odd are themselves odd functions. Indeed, this is the origin of the terms 'even function' and 'odd function'.

Exercise 7

(a) Is the function $C_0(t)$, given by $C_0(t) = 1$ for $-\tau/2 \le t \le \tau/2$, even, odd or neither?

(b) Is the function $h(t)$, defined by $h(t) = t^2 + t^3$ for all t, even, odd or neither?

The way that even and odd functions combine is similar to the way that positive and negative numbers combine. That is:

- the sum of two even functions (positive numbers) is even (positive)

- the sum of two odd functions (negative numbers) is odd (negative)

- the sum of an even function (positive number) and an odd function (negative number) is neither even nor odd (positive, negative or zero)

- the product of two even functions (positive numbers) is even (positive)

- the product of two odd functions (negative numbers) is even (positive)

- the product of an even function (positive number) and an odd function (negative number) is odd (negative).

In the next example and exercise, we use the first two properties in the list above to demonstrate the last two properties.

Example 3

If $f(t) = t^3 + 2t^5$ and $g(t) = t - t^3$, show that the function defined by $h(t) = f(t)\,g(t)$ is an even function.

Solution

Calculating explicitly,

$$
\begin{aligned}
h(t) = f(t)\,g(t) &= (t^3 + 2t^5)(t - t^3) \\
&= t^4 - t^6 + 2t^6 - 2t^8 \\
&= t^4 + t^6 - 2t^8.
\end{aligned}
$$

This is a polynomial where all the powers are even, therefore $h(t)$ is an even function.

Alternatively, since both $f(t)$ and $g(t)$ are odd functions, we know by definition that

$$
f(-t) = -f(t), \quad g(-t) = -g(t).
$$

Hence

$$
h(-t) = f(-t)\,g(-t) = (-f(t))(-g(t)) = f(t)\,g(t) = h(t),
$$

so $h(t)$ is an even function.

Exercise 8

If $f(t) = t^3 + 2t^5$ and $g(t) = 3t^2 - t^4$, show that the function defined by $h(t) = f(t)\,g(t)$ is an odd function.

For much of this subsection we have been concerned with even and odd functions that are not periodic. If a function $f(t)$ *is* periodic, of period $2a$, then an advantage of choosing a fundamental interval $[-a, a]$ centred on the origin is that we can tell whether $f(t)$ is even, odd or neither by seeing whether it is even, odd or neither on $[-a, a]$.

Even and odd periodic functions

Let $f(t)$ be periodic of period $2a$. Then $f(t)$ is even provided that it is even over the interval $[-a, a]$. Similarly, $f(t)$ is odd provided that it is odd over the interval $[-a, a]$.

Now we investigate the properties of odd and even functions that will simplify later calculations, namely what happens when these functions are integrated over the fundamental interval $[-a, a]$.

For an odd function $f(t)$, the integral is zero because the integral for positive t-values is exactly cancelled by the integral for negative t-values. This is illustrated in Figure 9 and is proved by the following argument. Calling the integral I and splitting it into two halves gives

$$I = \int_{-a}^{a} f(t)\,dt = \int_{-a}^{0} f(t)\,dt + \int_{0}^{a} f(t)\,dt.$$

Now use the rule that changing the order of integration changes the sign of the integral to obtain

$$I = -\int_{0}^{-a} f(t)\,dt + \int_{0}^{a} f(t)\,dt.$$

We can use the substitution $u = -t$ to yield

$$I = \int_{0}^{a} f(-u)\,du + \int_{0}^{a} f(t)\,dt.$$

Finally, use the fact that f is odd, so $f(-u) = -f(u)$, to get

$$I = -\int_{0}^{a} f(u)\,du + \int_{0}^{a} f(t)\,dt = 0,$$

where the final equality follows because the two integrals are the same integral written with different integration variables.

If f is an even function, then the argument above can be applied up until the penultimate step, then instead of terms cancelling, the result will be twice the integral from 0 to a.

These results are worth remembering as they can save a lot of effort when evaluating integrals of even and odd functions.

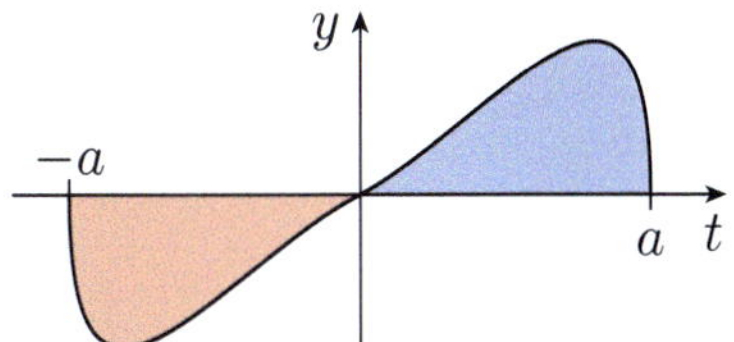

Figure 9   An odd function with area above the horizontal axis shaded blue and area below shaded red. By symmetry, the two areas are equal in size, so the integral is zero (as the red area is counted as negative).

Integrals of even and odd periodic functions

$$\int_{-a}^{a} g(t)\,dt = 2\int_{0}^{a} g(t)\,dt \quad \text{if } g \text{ is an even function.} \tag{7}$$

$$\int_{-a}^{a} f(t)\,dt = 0 \quad \text{if } f \text{ is an odd function.} \tag{8}$$

From Figure 7, you can see that the cosine function is even, but the sine function is odd. The suggestion is that to approximate an even function (such as the sawtooth function or the square-wave function) as a sum of sinusoidal terms, we ought to ensure that the approximating function is even. The only way to do this is to ensure that only cosine functions appear in the sum, as you will see in the next section.

2 Fourier series for even functions with period 2π

In this section you will see how to obtain the Fourier series for the sawtooth and square-wave functions, and in doing this you will obtain general formulas that can be applied to find the Fourier series for any even function with period 2π. We start with this special case as it is the simplest. However, even though it is a special case, you will see later that the arguments used to find the Fourier series also apply in the general case.

2.1 A series of approximations

In Section 1 you saw that there are certain useful things that can be said about the series $G(t)$ in equation (4), namely that it is periodic of period 2π and that it is an even function. But this does not tell us why this particular series corresponds to the sawtooth function $g(t)$ described in the Introduction. In this section we derive the Fourier series for the sawtooth function $g(t)$ by a mathematical argument, and we ask you to do the same for the square-wave function.

The graph of $g(t)$ (see Figure 10) coincides with the line $g(t) = 1 + t/\pi$ in the range $-\pi \le t < 0$ and with the line $g(t) = 1 - t/\pi$ in the range $0 \le t \le \pi$. The interval $[-\pi, \pi]$ is a fundamental interval for $g(t)$, on which it is defined as

$$g(t) = \begin{cases} \dfrac{1}{\pi}t + 1 & \text{for } -\pi \le t < 0, \\[2mm] -\dfrac{1}{\pi}t + 1 & \text{for } 0 \le t \le \pi. \end{cases} \tag{9}$$

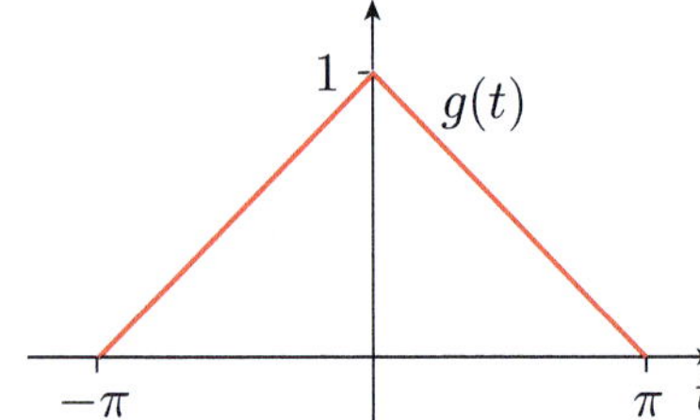

Figure 10 Graph of the sawtooth function $g(t)$

What are the period and angular frequency of the sawtooth function $g(t)$?

It seems reasonable that an even periodic function will have a Fourier series consisting of only even sinusoidal terms with the same period. This is indeed the case, as we will justify later. Here, we proceed to use a family of even sinusoidal functions, namely the family of cosine functions $\{C_0(t), C_1(t), C_2(t), \ldots\}$ that you met earlier:

$$C_n(t) = \cos nt \quad (n = 0, 1, 2, \ldots).$$

We consider a series

$$G(t) = A_0\, C_0(t) + A_1\, C_1(t) + A_2\, C_2(t) + A_3\, C_3(t) + \cdots$$

$$= A_0 + \sum_{n=1}^{\infty} A_n \cos nt, \tag{10}$$

and ask how the **Fourier coefficients** A_n can be chosen so that as we add successive terms of the series to obtain the approximations

We usually write the sum in this way with the constant term singled out because this form is easier to use in calculations.

$$G_0(t) = A_0,$$
$$G_1(t) = A_0 + A_1 \cos nt,$$
$$G_2(t) = A_0 + A_1 \cos nt + A_2 \cos(2nt),$$

and so on, the values approach $g(t)$ for any chosen value of t.

So the central problem is: how can A_n be chosen so that the approximation to $g(t)$ given by $G_n(t)$ gets better as n increases? We can also consider what happens when infinitely many terms are added to the approximation, so that we obtain the function $G(t)$. Can we choose the coefficients A_n so that $G(t) = g(t)$ for all t?

The argument that we use is quite general, in that it works for *any* even periodic function with fundamental interval $[-\pi, \pi]$. Thus for the remainder of this section, we will use the notation $f(t)$ to refer to a general such function, and

$$F(t) = A_0 + \sum_{n=1}^{\infty} A_n \cos nt \tag{11}$$

for the corresponding series whose coefficients $A_0, A_1, A_2, \ldots$ we are trying to find. We will then apply the general argument to the sawtooth function (in Examples 4–6), and ask you to apply it to other functions, including the square-wave function (in Exercises 11, 13 and 14).

We begin by deriving the first of the approximations listed above, namely $G_0(t)$.

2.2 A first approximation

The easiest coefficient to find is A_0. The technique is based on the observation that all the functions $C_n(t)$, some of which are illustrated in Figure 6 and the solution to Exercise 4, oscillate, and the positive contributions to an integral over the fundamental interval exactly cancel out the negative contributions. This means that if you integrate the cosine functions over the fundamental interval $[-\pi, \pi]$, then you obtain 0; that is,

$$\int_{-\pi}^{\pi} \cos nt \, dt = 0 \quad (n = 1, 2, 3, \dots). \tag{12}$$

Exercise 10

Verify that the integral in formula (12) is zero (for each $n = 1, 2, 3, \dots$).

In this module we assume the validity of term-by-term integration and differentiation of infinite series. The process is valid in all the practical cases that concern us.

If we integrate both sides of equation (11) term by term over the fundamental interval $[-\pi, \pi]$, then we find that

$$\int_{-\pi}^{\pi} F(t) \, dt = \int_{-\pi}^{\pi} \left(A_0 + \sum_{n=1}^{\infty} A_n \cos nt \right) dt$$

$$= \int_{-\pi}^{\pi} A_0 \, dt + \sum_{n=1}^{\infty} A_n \int_{-\pi}^{\pi} \cos nt \, dt.$$

Now all the terms involving integrals of cosine functions vanish, by formula (12), leaving us with

$$\int_{-\pi}^{\pi} F(t) \, dt = \int_{-\pi}^{\pi} A_0 \, dt = 2\pi A_0 \quad \text{(since } A_0 \text{ is a constant).}$$

Hence

$$A_0 = \frac{1}{2\pi} \int_{-\pi}^{\pi} F(t) \, dt.$$

We do not know the coefficients of $F(t)$, but our aim is to ensure that $F(t) = f(t)$. So to determine A_0, we replace $F(t)$ by $f(t)$ in the above equation to obtain the following result.

$$A_0 = \frac{1}{2\pi} \int_{-\pi}^{\pi} f(t) \, dt. \tag{13}$$

You can think of A_0 as the average value taken by the function $f(t)$.

Example 4

Find the value of A_0 when the general function $f(t)$ is replaced by the sawtooth function $g(t)$ given by equation (9).

Solution

Equation (13) becomes

$$A_0 = \frac{1}{2\pi} \int_{-\pi}^{\pi} g(t)\, dt.$$

As the function $g(t)$ is even, equation (7) applies and the integral is equal to twice the integral over the positive t-values:

$$A_0 = \frac{1}{\pi} \int_{0}^{\pi} g(t)\, dt.$$

Using this fact simplifies the calculation as $g(t)$ is defined piecewise, with different formulas for positive and negative values; here we need to consider only one of the formulas.

The function $g(t)$ is given by equation (9), so the above integral is

$$\frac{1}{\pi} \int_{0}^{\pi} g(t)\, dt = \frac{1}{\pi} \int_{0}^{\pi} \left(-\frac{1}{\pi} t + 1 \right) dt = \frac{1}{\pi} \left[-\frac{1}{2\pi} t^2 + t \right]_{0}^{\pi} = \frac{1}{2}.$$

Thus for the case of the sawtooth function,

$$A_0 = \tfrac{1}{2}.$$

The argument used to simplify the first step of the calculation in Example 4 is a general argument that applies whenever the integrand is an even function, and it could also be applied to equation (13) to obtain a simplified formula. But we will not do this here because, as we will see later, equation (13) is the formula that applies in general (not just to even functions).

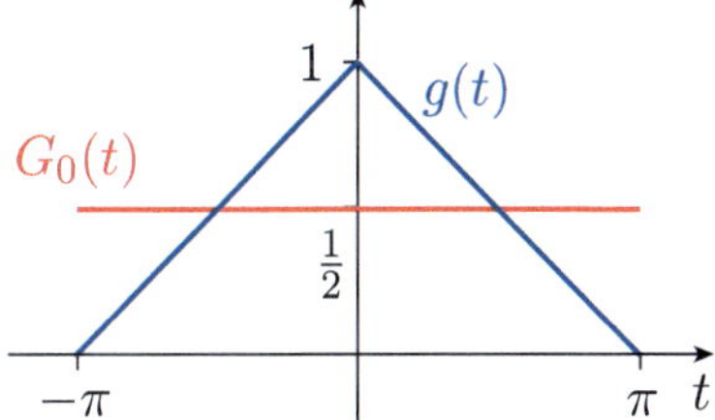

Figure 11   Approximation $G_0(t)$ compared to $g(t)$

Figure 11 shows graphs of the original function $g(t)$ and the first approximation $G_0(t) = A_0 = \tfrac{1}{2}$. You can see from the figure that $A_0 = \tfrac{1}{2}$ is the average value of the function $g(t)$.

Exercise 11

Suppose that the general function $f(t)$ is now replaced by the square-wave function $h(t)$ of the Introduction, which can be defined on the fundamental interval $[-\pi, \pi]$ by

$$h(t) = \begin{cases} 1 & \text{for } -\frac{\pi}{2} \le t \le \frac{\pi}{2}, \\ 0 & \text{otherwise.} \end{cases}$$

Find the value of the Fourier coefficient A_0.

Hint: As the function $h(t)$ is zero over some of the fundamental interval,

$$\int_{-\pi}^{\pi} h(t)\, dt = \int_{-\pi/2}^{\pi/2} h(t)\, dt.$$

Integration enabled us to eliminate the coefficients A_n $(n > 0)$ and hence to find the coefficient A_0. The next subsection looks at how we can find the coefficient A_1 and hence find a better approximation to $g(t)$.

2.3 A second approximation

To find the next coefficient, A_1, we must somehow eliminate A_0 and all the other coefficients. The technique is to multiply both sides of equation (11) by the term $\cos t$ to give

$$F(t)\cos t = A_0 \cos t + \sum_{n=1}^{\infty} A_n \cos nt \,\cos t. \tag{14}$$

If we integrate both sides of equation (14) term by term over the fundamental interval $[-\pi, \pi]$, then we find that

$$\int_{-\pi}^{\pi} F(t)\cos t\, dt = \int_{-\pi}^{\pi} A_0 \cos t\, dt + \sum_{n=1}^{\infty}\left(A_n \int_{-\pi}^{\pi} \cos nt \cos t\, dt\right). \tag{15}$$

Now we evaluate each of these integrals separately.

Exercise 12

(a) Show that

$$\int_{-\pi}^{\pi} A_0 \cos t\, dt = 0.$$

(b) Use the trigonometric identity

$$\cos\alpha \cos\beta = \tfrac{1}{2}\cos(\alpha+\beta) + \tfrac{1}{2}\cos(\alpha-\beta)$$

to show that

$$\int_{-\pi}^{\pi} \cos 2t \cos t\, dt = 0.$$

(c) More generally, use the identity given in part (b) to show that

$$\int_{-\pi}^{\pi} \cos nt \cos t\, dt = 0 \quad \text{when } n \text{ is an integer and } n > 1.$$

(d) Use the trigonometric identity

$$\cos 2\alpha = 2\cos^2\alpha - 1$$

to evaluate the integral

$$\int_{-\pi}^{\pi} \cos^2 t\, dt.$$

Exercise 12 shows that most of the integrals on the right-hand side of equation (15) evaluate to zero. The only remaining term involves the coefficient A_1, and we are left with

$$\int_{-\pi}^{\pi} F(t)\cos t\, dt = A_1 \int_{-\pi}^{\pi} \cos^2 t\, dt = A_1 \pi.$$

Hence the coefficient A_1 is given by

$$A_1 = \frac{1}{\pi}\int_{-\pi}^{\pi} F(t)\cos t\, dt.$$

We are trying to choose the coefficients so that $F(t) = f(t)$, so we replace $F(t)$ by $f(t)$ and obtain the following result.

$$A_1 = \frac{1}{\pi} \int_{-\pi}^{\pi} f(t) \cos t \, dt. \tag{16}$$

Before applying this formula to calculate A_1 for the sawtooth function, we state two useful integrals that often arise when calculating Fourier series.

Two useful integrals

For a a non-zero constant and C a constant,

$$\int t \sin(at) \, dt = \frac{1}{a^2} \left(\sin(at) - at \cos(at) \right) + C, \tag{17}$$

$$\int t \cos(at) \, dt = \frac{1}{a^2} \left(\cos(at) + at \sin(at) \right) + C. \tag{18}$$

Both of these integrals are easy to derive using integration by parts, but it is quicker to state the standard result. One of these integrals will be used in the calculation of A_1 for the sawtooth function in the next example.

Example 5

Returning to the sawtooth function $g(t)$ defined by equation (9), find the value of the coefficient A_1.

Solution

The coefficient is given by

$$A_1 = \frac{1}{\pi} \int_{-\pi}^{\pi} g(t) \cos t \, dt.$$

As $g(t)$ is even and $\cos t$ is even, the product $g(t) \cos t$ is even, and we may make use of equation (7) to simplify the calculation by evaluating twice the integral over the positive t-values:

$$A_1 = \frac{2}{\pi} \int_{0}^{\pi} g(t) \cos t \, dt.$$

Substituting the definition of $g(t)$ from equation (9) gives

$$A_1 = \frac{2}{\pi} \int_{0}^{\pi} \left(-\frac{1}{\pi} t + 1 \right) \cos t \, dt$$

$$= -\frac{2}{\pi^2} \int_{0}^{\pi} t \cos t \, dt + \frac{2}{\pi} \int_{0}^{\pi} \cos t \, dt.$$

The first integral is one of the two useful integrals (equation (18) with $a = 1$) that were stated immediately preceding this example, so performing the integrations yields

$$A_1 = -\frac{2}{\pi^2}\big[\cos t + t\sin t\big]_0^\pi + \frac{2}{\pi}\big[\sin t\big]_0^\pi$$

$$= -\frac{2}{\pi^2}(-1-1) = \frac{4}{\pi^2}.$$

So our second approximation to the function $g(t)$ (our first non-constant approximation) is

$$G_1(t) = \frac{1}{2} + \frac{4}{\pi^2}\cos t.$$

The graph of this approximation is compared with the graph of $g(t)$ in Figure 12. Already $G_1(t)$ is quite a reasonable approximation to $g(t)$.

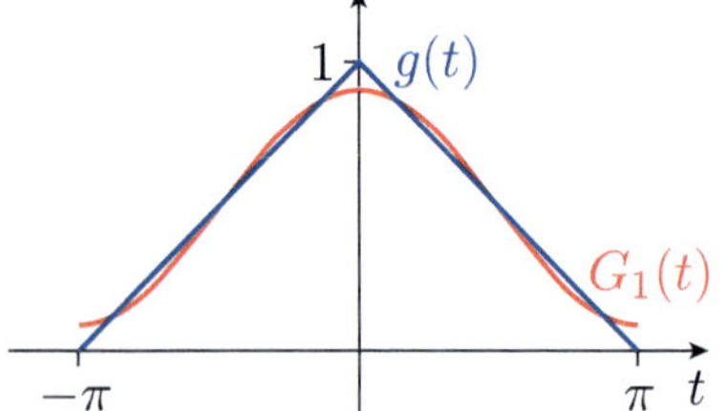

Figure 12 Approximation $G_1(t)$ compared to $g(t)$

Exercise 13

Suppose that $f(t)$ in equation (16) is replaced by the square-wave function $h(t)$ of the Introduction, defined in Exercise 11. Find the value of the coefficient A_1.

2.4 Better approximations

The coefficient A_n can be found in a similar way to the way in which we found A_1, by multiplying by an appropriate sinusoidal function and integrating. In order to do this, we need to evaluate many integrals, but like those in Exercise 12, most vanish. Generalising the results of that exercise, we find the following.

> ### Trigonometric integrals over the interval $[-\pi, \pi]$
>
> For any positive integers m and n,
>
> $$\int_{-\pi}^{\pi} \cos mt \, dt = 0, \tag{19}$$
>
> $$\int_{-\pi}^{\pi} \cos mt \cos nt \, dt = 0 \quad (m \neq n), \tag{20}$$
>
> $$\int_{-\pi}^{\pi} \cos^2 mt \, dt = \pi. \tag{21}$$

These results mean that if we multiply both sides of equation (11) by $\cos mt$ and integrate over the fundamental interval, then all of the coefficients except A_m disappear. To see this, multiply both sides of equation (11) by $\cos mt$, to obtain

$$F(t)\cos mt = A_0 \cos mt + \sum_{n=1}^{\infty} A_n \cos mt \, \cos nt.$$

Then integration gives

$$\int_{-\pi}^{\pi} F(t)\cos mt \, dt = \int_{-\pi}^{\pi} A_0 \cos mt \, dt + \sum_{n=1}^{\infty} A_n \int_{-\pi}^{\pi} \cos mt \, \cos nt \, dt,$$

and using formulas (19) and (20), we find that all of the terms of the above infinite sum are zero except when $n = m$, so we get

$$\int_{-\pi}^{\pi} F(t)\cos mt \, dt = A_m \int_{-\pi}^{\pi} \cos^2 mt \, dt.$$

Finally, formula (21) gives the value of the right-hand integral as π, so

$$A_m = \frac{1}{\pi} \int_{-\pi}^{\pi} F(t)\cos mt \, dt.$$

This key result is worth remembering, so we re-state it in the form with the original function that we are trying to approximate, $f(t)$, instead of $F(t)$, and the usual index n instead of m.

$$A_n = \frac{1}{\pi} \int_{-\pi}^{\pi} f(t)\cos nt \, dt \quad (n > 0). \tag{22}$$

Example 6

Returning once again to the sawtooth function $g(t)$ as defined by equation (9), find the values of the coefficients A_2 and A_3.

Solution

Substituting $g(t)$ into equation (22) gives

$$A_n = \frac{1}{\pi} \int_{-\pi}^{\pi} g(t)\cos nt \, dt.$$

Now we use the fact that the integrand is even, since it is the product of two even functions, $g(t)$ and $\cos nt$, to obtain

$$A_n = \frac{2}{\pi} \int_{0}^{\pi} \left(-\frac{1}{\pi}t + 1\right) \cos nt \, dt$$

$$= -\frac{2}{\pi^2} \int_{0}^{\pi} t \cos nt \, dt + \frac{2}{\pi} \int_{0}^{\pi} \cos nt \, dt,$$

where we have substituted for $g(t)$ using the definition in equation (9).

This integral can be evaluated by recognising the first integral as one of the two useful integrals (equation (18) with $a = n$), to give

$$A_n = -\frac{2}{\pi^2}\left[\frac{1}{n^2}(\cos nt + nt \sin nt)\right]_0^\pi + \frac{2}{\pi}\left[\frac{1}{n}\sin nt\right]_0^\pi$$

$$= -\frac{2}{\pi^2}\left(\frac{1}{n^2}((-1)^n - 1)\right),$$

since $\sin n\pi = 0$ and $\cos n\pi = (-1)^n$, so

$$A_n = \frac{2}{n^2\pi^2}(1 - (-1)^n).$$

If $n = 2$, then $(-1)^n = 1$, so $A_2 = 0$.

If $n = 3$, then $(-1)^n = -1$, so $A_3 = 4/(9\pi^2)$.

From Example 6 we have $A_2 = 0$, so the approximation $G_2(t)$ is equal to $G_1(t)$ and hence is no better as an approximation to $g(t)$. However, from Examples 4, 5 and 6, we can derive a better approximation to $g(t)$ by writing

$$G_3(t) = A_0 + A_1 \cos t + A_2 \cos 2t + A_3 \cos 3t$$

$$= \frac{1}{2} + \frac{4}{\pi^2}\left(\cos t + \frac{1}{9}\cos 3t\right).$$

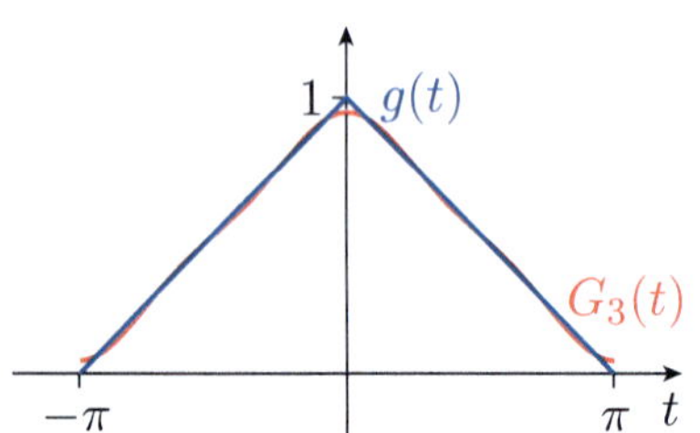

Figure 13 Approximation $G_3(t)$ compared to $g(t)$

The graph of $G_3(t)$ is shown in Figure 13. This shows a further improvement in the accuracy of the approximation, to the extent that on this scale it is hard to see a difference between the graphs of $G_3(t)$ and $g(t)$.

Exercise 14

Suppose that $f(t)$ in formula (22) is replaced by the square-wave function $h(t)$, defined in Exercise 11. Find the values of the coefficients A_2 and A_3.

Examples 4, 5 and 6 can be generalised to find all the coefficients in the Fourier series for the sawtooth function $g(t)$. We find that

$$A_0 = \frac{1}{2}, \quad A_1 = \frac{4}{\pi^2}, \quad A_2 = 0, \quad A_3 = \frac{4}{\pi^2} \times \frac{1}{9}, \quad A_4 = 0,$$

$$A_5 = \frac{4}{\pi^2} \times \frac{1}{25}, \quad A_6 = 0, \quad A_7 = \frac{4}{\pi^2} \times \frac{1}{49}, \quad A_8 = 0, \quad \dots,$$

and a clear pattern has appeared. Hence the Fourier series for the sawtooth function $g(t)$ is

$$G(t) = \frac{1}{2} + \frac{4}{\pi^2}\left(\cos t + \frac{1}{9}\cos 3t + \frac{1}{25}\cos 5t + \frac{1}{49}\cos 7t + \cdots\right),$$

confirming equation (2).

The terms of this Fourier series can be written more compactly using the sigma notation for summations. The angular frequencies of the terms form a pattern of successive odd numbers. The coefficient of each term in the brackets is the reciprocal of the square of the angular frequency. Recall that if $n = 1, 2, 3, \ldots$, then $2n = 2, 4, 6, \ldots$ runs through the even numbers and $2n - 1 = 1, 3, 5, \ldots$ runs through the odd numbers. With these observations, the Fourier series can be written as

$$G(t) = \frac{1}{2} + \frac{4}{\pi^2} \sum_{n=1}^{\infty} \frac{1}{(2n-1)^2} \cos(2n-1)t.$$

It is sometimes more convenient to write Fourier series in so-called **closed form** like this, using the summation symbol $\sum$. Now try doing this yourself.

Exercise 15

You have found, in Exercises 11, 13 and 14, the Fourier coefficients A_0, A_1, A_2 and A_3 for the square-wave function $h(t)$ defined in Exercise 11. In fact, as indicated in equation (3), its Fourier series is

$$H(t) = \frac{1}{2} + \frac{2}{\pi} \cos t - \frac{2}{3\pi} \cos 3t + \frac{2}{5\pi} \cos 5t - \frac{2}{7\pi} \cos 7t + \cdots .$$

Write down this series in closed form.

The following exercises ask you to apply the formulas derived in this section to find Fourier series for other even periodic functions.

Exercise 16

Find the Fourier series for the even periodic function $f(t)$ defined on the fundamental interval $[-\pi, \pi]$ by

$$f(t) = t^2.$$

(*Hint*: You may find the following integral obtained by integration by parts useful:

$$\int t^2 \cos nt \, dt = \frac{1}{n^3} \left((n^2 t^2 - 2) \sin nt + 2nt \cos nt \right) + C,$$

where C is a constant.)

Exercise 17

A variant of the sawtooth function can be defined on the fundamental interval $[-\pi, \pi]$ by

$$w(t) = |t|.$$

Recall that the absolute value function is defined by

$$|t| = \begin{cases} t & \text{for } t \geq 0, \\ -t & \text{for } t < 0. \end{cases}$$

Find the Fourier series for this function, and write down the approximation $W_5(t)$.

2.5 Convergence

You have just tackled a substantial piece of work, and this has involved finding Fourier series for several even functions. Having found them, you need to take stock of what you have done. You have seen that just a few terms of the Fourier series for the sawtooth function $g(t)$ give a very good approximation. However, the first few terms of the Fourier series for the square-wave function $h(t)$ do not give a particularly good approximation, as Figure 14 illustrates.

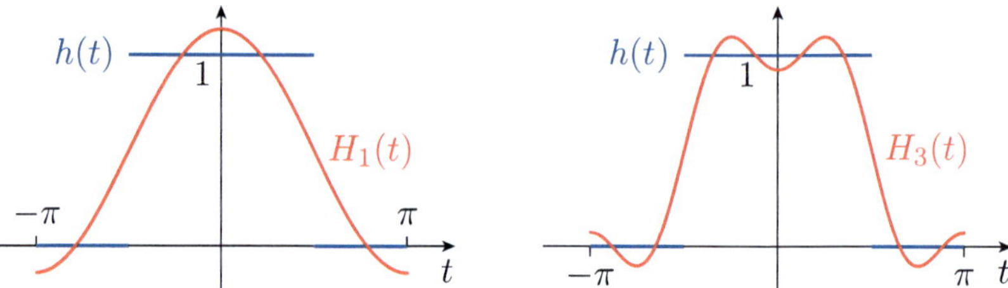

Figure 14   The first two approximations (red lines) to the square-wave function (blue line) by its Fourier series

The situation improves only slowly for the square-wave function. Plotting the graphs for the sums as far as the $\cos 7t$, $\cos 11t$ and $\cos 21t$ terms, better approximations to $h(t)$ are obtained, as expected (see Figure 15).

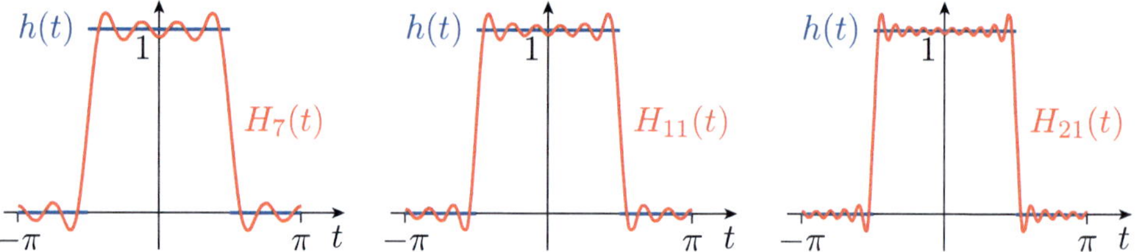

Figure 15   More approximations to the square-wave function by its Fourier series

However, even $H_{21}(t)$ does not approximate $h(t)$ as well as $G_3(t)$ does $g(t)$ in Figure 13. This is because of the discontinuities in $h(t)$. We cannot reasonably expect the sum of continuous sinusoidal functions to provide a good approximation to a discontinuous function. From the graphs, you can see that the approximations to $h(t)$ are worse near the discontinuities, that is, near the points where the value of $h(t)$ jumps from 0 to 1 and back again. Nevertheless, even for a discontinuous function such as $h(t)$, we can, remarkably, approximate reasonably well using Fourier series. At a discontinuity, the Fourier series takes the average value of the function at either side of the discontinuity. This is formally stated in the following theorem (which we do not prove) that guarantees the nature of the Fourier series for a wide class of functions at points in the fundamental interval.

> ### Theorem 1   Pointwise convergence
>
> If, on the interval $[-\pi, \pi]$, the function f has a continuous derivative except at a finite number of points, then at each point $x_0 \in [-\pi, \pi]$, the Fourier series for f converges to
>
> $$\tfrac{1}{2}\left(f(x_0^+) + f(x_0^-)\right).$$
>
> Here $f(x_0^+)$ is the limit of $f(x)$ as x approaches x_0 from above, and $f(x_0^-)$ is the limit of $f(x)$ as x approaches x_0 from below.

It is worth remarking that if f is continuous at x_0, then $f(x_0^-) = f(x_0^+) = f(x_0)$, and in this case the theorem states that the Fourier series converges to $f(x_0)$.

The following example shows how Theorem 1 is used to determine the values to which Fourier series converge.

Example 7

Consider the function

$$f(t) = \begin{cases} -1 & \text{for } -1 \le t < 0, \\ t & \text{for } 0 \le t < 1, \end{cases}$$

$$f(t + 2) = f(t),$$

and its corresponding Fourier series $F(t)$.

(a) Calculate $f(-1)$, $f(\tfrac{1}{2})$ and $f(2)$.

(b) Calculate $F(-1)$, $F(\tfrac{1}{2})$ and $F(2)$.

(c) Compare the values obtained in parts (a) and (b).

Solution

(a) The first step in solving problems such as this is to draw a sketch graph, such as the one shown in Figure 16.

Using the sketch as a guide, we can calculate

$$f(-1) = -1,$$
$$f(\tfrac{1}{2}) = \tfrac{1}{2},$$
$$f(2) = f(0) = 0.$$

(b) The pointwise convergence theorem gives

$$F(-1) = \frac{f(-1^+) + f(-1^-)}{2}.$$

Approaching $t = -1$ from above is included in the fundamental interval of f, so we can say that $f(-1^+) = -1$.

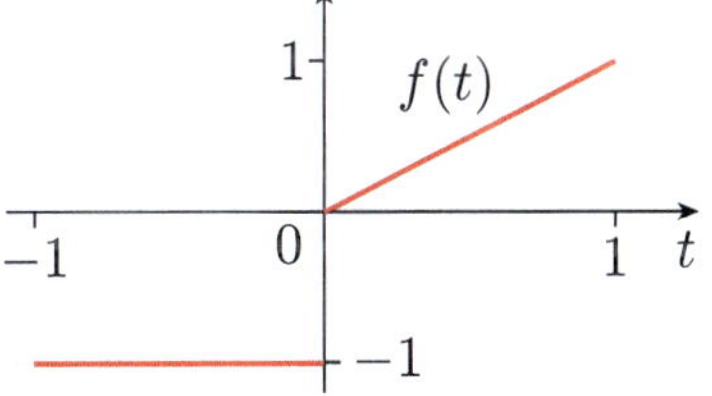

Figure 16   Sketch graph of $f(t)$ on its fundamental interval

Approaching $t = -1$ from below is not in the fundamental interval, so we use the periodicity of f to say that the value is the same as approaching $t = 1$ from below, that is, $f(-1^-) = f(1^-)$. So from the sketch we have $f(1^-) = 1$.

Substituting these values into the formula gives

$$F(-1) = \frac{(-1) + 1}{2} = 0.$$

At $t = \frac{1}{2}$ the function $f(t)$ is continuous, so the Fourier series will be equal to the given function here, hence

$$F(\tfrac{1}{2}) = f(\tfrac{1}{2}) = \tfrac{1}{2}.$$

As F is also periodic with period 2, we have $F(2) = F(0)$. So

$$F(2) = F(0) = \frac{f(0^+) + f(0^-)}{2}.$$

Now, using the sketch of the function as a guide, we obtain that approaching $t = 0$ from above gives $f(0^+) = 0$ and approaching $t = 0$ from below gives $f(0^-) = -1$.

Substituting these values into the formula gives

$$F(2) = \frac{0 + (-1)}{2} = -\tfrac{1}{2}.$$

(c) We have $f(\tfrac{1}{2}) = F(\tfrac{1}{2})$, which is to be expected as f is continuous at $t = \frac{1}{2}$.

If f is not continuous at a point, then the Fourier series does not necessarily converge to the value of the function. This is the case for the other two points considered here, where we have found that $f(-1) \neq F(-1)$ and $f(2) \neq F(2)$. Both of these points are points of discontinuity of f.

Now try the following exercise in applying the pointwise convergence theorem.

Exercise 18

Consider the function

$$f(t) = \frac{|t| + t}{2} \quad \text{for } -1 \leq t < 1,$$

$$f(t + 2) = f(t),$$

and its corresponding Fourier series $F(t)$. Calculate the values $F(-1)$ and $F(0)$.

3 Fourier series for even and odd periodic functions

In the previous section we concentrated on finding the Fourier series for two particular even periodic functions with period $\tau = 2\pi$ and hence fundamental interval $[-\pi, \pi]$. In this section we extend the technique to periodic functions that are either even or odd and have any fixed period τ.

3.1 Fourier series for even functions with period τ

We will now examine a generalisation of the function $w(t)$ described in Exercise 17 that has period τ, where τ is a specific positive number. The graph of $w(t)$ repeats along the t-axis and looks like a sawtooth function (see Figure 17).

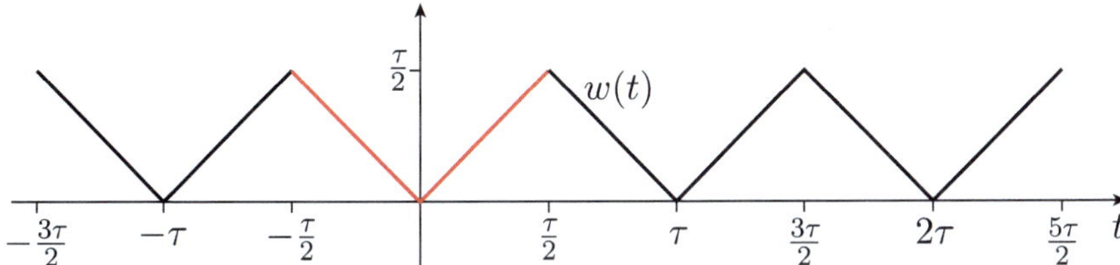

Figure 17   Graph of the function $w(t)$ over four periods, with the fundamental interval highlighted in red

If the function is defined to have period τ (Exercise 17 considered the special case $\tau = 2\pi$), then a fundamental interval is $\left[-\frac{\tau}{2}, \frac{\tau}{2}\right]$, on which $w(t)$ is shown as the red curve in Figure 17. On this interval the function is given by $w(t) = |t|$ or

$$w(t) = \begin{cases} -t & \text{for } -\frac{\tau}{2} \leq t < 0, \\ t & \text{for } 0 \leq t < \frac{\tau}{2}. \end{cases} \tag{23}$$

You saw in Subsection 1.1 that when we are trying to find a Fourier series for a function with period τ, we must also consider functions with the shorter periods

$$\frac{\tau}{2}, \frac{\tau}{3}, \frac{\tau}{4}, \frac{\tau}{5}, \ldots$$

Corresponding to the fundamental period τ and to these shorter periods are the angular frequencies

$$\frac{2\pi}{\tau}, \frac{4\pi}{\tau}, \frac{6\pi}{\tau}, \frac{8\pi}{\tau}, \frac{10\pi}{\tau}, \ldots$$

Hence we consider the family of even functions

$$C_n(t) = \cos\left(\frac{2n\pi t}{\tau}\right) \quad (n = 1, 2, 3, \ldots).$$

Since we are dealing with even functions, we also include the constant function

$$C_0(t) = 1.$$

As in Section 2, the argument used here is a general one. Thus for the remainder of this subsection, we use the symbol $f(t)$ to refer to a general even periodic function with fundamental interval $\left[-\frac{\tau}{2}, \frac{\tau}{2}\right]$. As before, we assume that we can choose the coefficients in an infinite sum of these functions $C_n(t)$ in such a way that we can approximate the original function as accurately as required. We write this (as before) as

$$F(t) = A_0 + \sum_{n=1}^{\infty} A_n \cos\left(\frac{2n\pi t}{\tau}\right). \tag{24}$$

To find the coefficients in this sum, we need to evaluate integrals of the cosine functions that generalise equations (19)–(21) used in the previous section. The integrals in which we are interested are for functions defined over the fundamental interval $\left[-\frac{\tau}{2}, \frac{\tau}{2}\right]$, which are obtained from the former integrals by the substitution $u = \tau t/(2\pi)$.

> **Trigonometric integrals over the interval $\left[-\frac{\tau}{2}, \frac{\tau}{2}\right]$**
>
> For any positive integers m and n,
>
> $$\int_{-\tau/2}^{\tau/2} \cos\left(\frac{2m\pi t}{\tau}\right) dt = 0, \tag{25}$$
>
> $$\int_{-\tau/2}^{\tau/2} \cos\left(\frac{2m\pi t}{\tau}\right) \cos\left(\frac{2n\pi t}{\tau}\right) dt = 0 \quad (m \neq n), \tag{26}$$
>
> $$\int_{-\tau/2}^{\tau/2} \cos^2\left(\frac{2m\pi t}{\tau}\right) dt = \frac{\tau}{2}. \tag{27}$$

To find the coefficients in Fourier series (24), we proceed as before. We first multiply both sides of equation (24) by a chosen cosine function. Then we integrate, and all but one of the coefficients become zero. We are left with a formula for that remaining coefficient.

First, to find the constant A_0, we integrate both sides of equation (24) over the fundamental interval $\left[-\frac{\tau}{2}, \frac{\tau}{2}\right]$ to obtain

$$\int_{-\tau/2}^{\tau/2} F(t)\, dt = \int_{-\tau/2}^{\tau/2} A_0\, dt + \sum_{n=1}^{\infty} A_n \int_{-\tau/2}^{\tau/2} \cos\left(\frac{2n\pi t}{\tau}\right) dt.$$

Using formula (25), all the integrals in the infinite sum become zero, leaving

$$\int_{-\tau/2}^{\tau/2} F(t)\, dt = \int_{-\tau/2}^{\tau/2} A_0\, dt = \tau A_0,$$

so

$$A_0 = \frac{1}{\tau} \int_{-\tau/2}^{\tau/2} F(t)\, dt.$$

As in Section 2, we now use the fact that we wish to choose the coefficients so that $F(t) = f(t)$. Thus we put $F(t) = f(t)$ to give the following result.

$$A_0 = \frac{1}{\tau} \int_{-\tau/2}^{\tau/2} f(t)\, dt. \tag{28}$$

The constant A_0 can again be thought of as the average value of the function $f(t)$ on the fundamental interval.

The derivation of a formula for the coefficient A_n for functions with period τ is very similar to the derivation of equation (22) for functions with period 2π. We will not repeat the argument; we simply state the result.

$$A_n = \frac{2}{\tau} \int_{-\tau/2}^{\tau/2} f(t) \cos\left(\frac{2n\pi t}{\tau}\right) dt. \tag{29}$$

Now apply these results to do the following exercise.

Exercise 19

For the sawtooth function $w(t)$ defined by equation (23), find the coefficients A_0 and A_n.

Substituting the coefficients that you have found in Exercise 19 into equation (24) gives the Fourier series $W(t)$ corresponding to the sawtooth function $w(t)$:

$$
\begin{aligned}
W(t) &= \frac{\tau}{4} - \frac{\tau}{\pi^2} \sum_{n=1}^{\infty} \frac{(-1)^n - 1}{n^2} \cos\left(\frac{2n\pi t}{\tau}\right) \\
&= \frac{\tau}{4} - \frac{\tau}{\pi^2}\left(2\cos\left(\frac{2\pi t}{\tau}\right) + \frac{2}{9}\cos\left(\frac{6\pi t}{\tau}\right)\right. \\
&\qquad\qquad \left. + \frac{2}{25}\cos\left(\frac{10\pi t}{\tau}\right) + \cdots \right).
\end{aligned}
\tag{30}
$$

Since $w(t)$ is very similar in form to the sawtooth function that we investigated in Section 2, it should come as no surprise that the successive approximations to $w(t)$ generated by series (30) converge rapidly to $w(t)$. Figure 18 compares the graph of $w(t)$ with the graphs of the approximations

$$W_0(t) = \frac{\tau}{4} \quad \text{and} \quad W_1(t) = \frac{\tau}{4} - \frac{2\tau}{\pi^2}\cos\left(\frac{2\pi t}{\tau}\right).$$

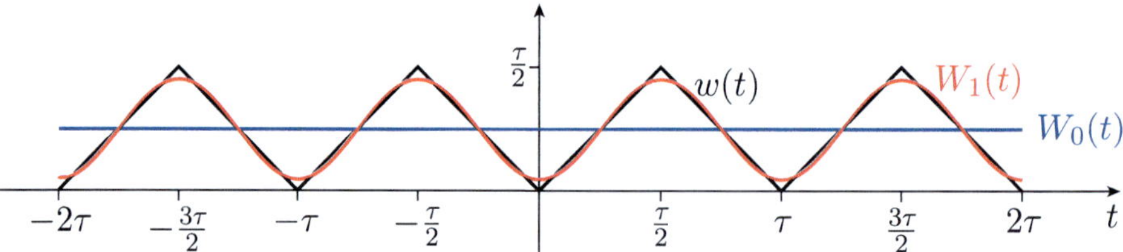

Figure 18   The sawtooth function $w(t)$ compared to the first two Fourier series approximations

3.2 Fourier series for odd functions with period τ

We have so far concentrated on even periodic functions, but it is equally straightforward to deal with odd periodic functions. As an example of an odd periodic function, consider the function $v(t)$ whose graph is shown in Figure 19. You can think of $v(t)$ as representing another type of sawtooth function, or as a broken surface made up of successive ramps and steps.

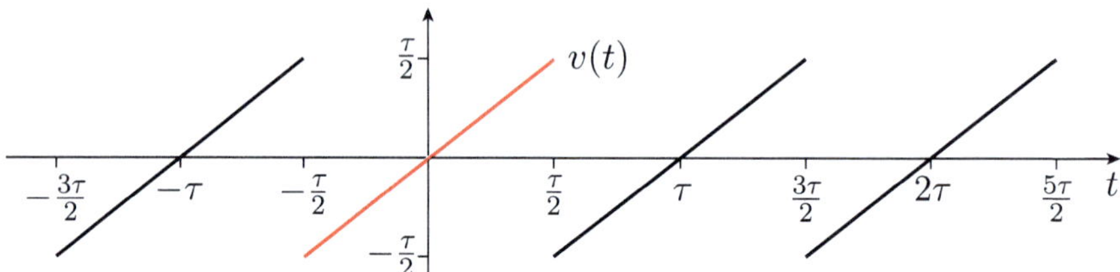

Figure 19   Graph of the function $v(t)$ with the fundamental interval highlighted in red

The function $v(t)$ is defined to be equal to t within the fundamental interval $-\tau/2 < t < \tau/2$, but what values should be assigned to the ends of the fundamental interval, that is, at $t = -\tau/2$ and $t = \tau/2$? Note that if the function has period τ, then $v(t) = v(t + \tau)$, so with $t = -\tau/2$ we have $v(-\tau/2) = v(\tau + (-\tau/2)) = v(\tau/2)$. For our purposes it makes no difference how $v(t)$ is defined on these points because we are only interested in integrals over the fundamental interval of $v(t)$, and changing values at a single point does not affect the integrals. As we are free to choose, we conventionally pick the value at the left of the interval and define $v(t)$ as

$$v(t) = t \quad \text{for } -\tau/2 \le t < \tau/2, \tag{31}$$
$$v(t + \tau) = v(t).$$

The second line of this definition repeats the definition on the fundamental interval to make $v(t)$ have period τ.

The function $v(t)$ as defined above is essentially an odd function as it differs from an odd function only at the endpoints (an odd periodic function is zero at the endpoints). For our purposes this is close enough to being an odd function as changing a function at a single point does not change the value of an integral of the function.

In order to approximate the odd function $v(t)$, we need odd trigonometric functions with period τ. As usual, we must consider functions with periods

$$\tau,\ \frac{\tau}{2},\ \frac{\tau}{3},\ \frac{\tau}{4},\ \frac{\tau}{5},\ \dots\ .$$

Corresponding to these periods are the angular frequencies

$$\frac{2\pi}{\tau},\ \frac{4\pi}{\tau},\ \frac{6\pi}{\tau},\ \frac{8\pi}{\tau},\ \frac{10\pi}{\tau},\ \dots\ .$$

So we must consider the family of sine functions

$$S_n(t) = \sin\left(\frac{2n\pi t}{\tau}\right) \quad (n = 1, 2, 3, \dots),$$

which is a family of odd functions.

In contrast to Subsection 3.1, where we considered the cosine series, we do not bother with the function $S_0(t) = \sin 0 = 0$, since any multiple of this function is zero.

Example 8

Sketch the graphs of the functions $S_1(t)$, $S_2(t)$, $S_3(t)$ and $S_4(t)$ on the fundamental interval $\left[-\frac{\tau}{2}, \frac{\tau}{2}\right]$.

Solution

Sketches of the four functions are shown in Figure 20.

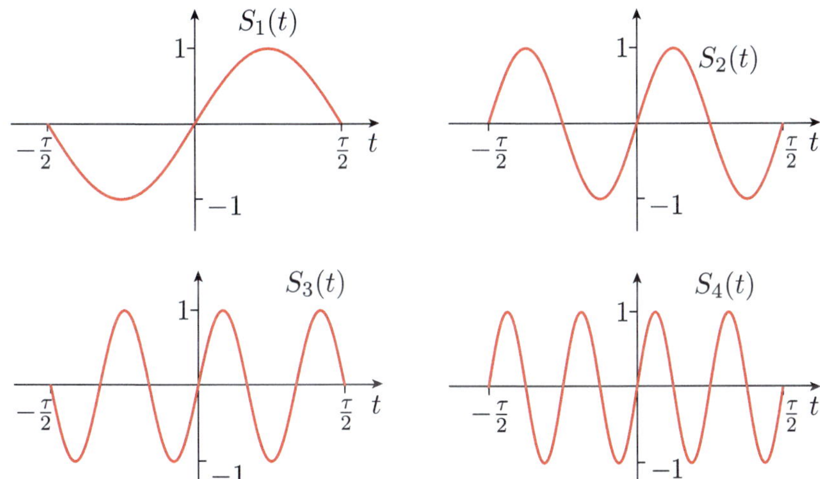

Figure 20 The first four members of a family of sine functions

As before, the argument used to find the coefficients in a Fourier series for an odd function is a general one. Thus for the remainder of this subsection we use the function $f(t)$ to refer to a general odd periodic function with fundamental interval $\left[-\frac{\tau}{2}, \frac{\tau}{2}\right]$. Then we assume that we can choose the coefficients in an infinite sum of the functions $S_n(t)$ in such a way that we can approximate the original function $f(t)$ as accurately as required.

We write the sum as

$$F(t) = \sum_{n=1}^{\infty} B_n \sin\left(\frac{2n\pi t}{\tau}\right), \tag{32}$$

where the coefficients $B_1, B_2, B_3, \ldots$ are constants depending on the particular function $f(t)$. We will refer to equation (32) as the Fourier series for the odd function $f(t)$.

To find the coefficients B_n, we multiply both sides of equation (32) by the function $\sin(2m\pi t/\tau)$ and integrate over the fundamental interval to give

$$\int_{-\tau/2}^{\tau/2} F(t) \sin\left(\frac{2m\pi t}{\tau}\right) dt$$

$$= \sum_{n=1}^{\infty} B_n \int_{-\tau/2}^{\tau/2} \sin\left(\frac{2m\pi t}{\tau}\right) \sin\left(\frac{2n\pi t}{\tau}\right) dt. \tag{33}$$

To simplify equation (33), we need evaluated integrals of sine functions analogous to those for the cosine functions in Subsection 3.1.

Trigonometric integrals over the interval $\left[-\frac{\tau}{2}, \frac{\tau}{2}\right]$

For any positive integers m and n,

$$\int_{-\tau/2}^{\tau/2} \sin\left(\frac{2m\pi t}{\tau}\right) dt = 0, \tag{34}$$

$$\int_{-\tau/2}^{\tau/2} \sin\left(\frac{2m\pi t}{\tau}\right) \sin\left(\frac{2n\pi t}{\tau}\right) dt = 0 \quad (m \neq n), \tag{35}$$

$$\int_{-\tau/2}^{\tau/2} \sin^2\left(\frac{2m\pi t}{\tau}\right) dt = \frac{\tau}{2}. \tag{36}$$

Using formula (35), all the integrals on the right-hand side of equation (33) become zero except when $n = m$. That is,

$$\int_{-\tau/2}^{\tau/2} F(t) \sin\left(\frac{2m\pi t}{\tau}\right) dt = B_m \int_{-\tau/2}^{\tau/2} \sin^2\left(\frac{2m\pi t}{\tau}\right) dt.$$

Using formula (36), the right-hand side reduces to $B_m \tau/2$, and we can make B_m the subject (and as before replace $F(t)$ by the function $f(t)$ and rewrite the index as n instead of m).

$$B_n = \frac{2}{\tau} \int_{-\tau/2}^{\tau/2} f(t) \sin\left(\frac{2n\pi t}{\tau}\right) dt. \tag{37}$$

Example 9

(a) Find the coefficients B_1, B_2 and B_3 for the function $v(t)$ defined in equation (31).

(b) Sketch the graph of

$$V_3(t) = B_1 \sin\left(\frac{2\pi t}{\tau}\right) + B_2 \sin\left(\frac{4\pi t}{\tau}\right) + B_3 \sin\left(\frac{6\pi t}{\tau}\right)$$

on the interval $\left[-\frac{\tau}{2}, \frac{\tau}{2}\right]$, and compare it with the graph of $v(t)$.

Solution

(a) The coefficients B_n are given by formula (37) with $f(t) = t$:

$$B_n = \frac{2}{\tau} \int_{-\tau/2}^{\tau/2} t \sin\left(\frac{2n\pi t}{\tau}\right) dt.$$

As the function $f(t) = t$ is odd and the sine term is odd, the product is even, so we can use equation (7) to rewrite the integral as twice the sum over the positive t-values:

$$B_n = \frac{4}{\tau} \int_0^{\tau/2} t \sin\left(\frac{2n\pi t}{\tau}\right) dt.$$

This integral is one of the two useful integrals (equation (17) with $a = 2n\pi/\tau$), so

$$B_n = \frac{4}{\tau} \left[\frac{\tau^2}{4n^2\pi^2} \left(\sin\left(\frac{2n\pi t}{\tau}\right) - \frac{2n\pi t}{\tau} \cos\left(\frac{2n\pi t}{\tau}\right) \right) \right]_0^{\tau/2}$$

$$= \frac{\tau}{n^2\pi^2} \left[\sin\left(\frac{2n\pi t}{\tau}\right) - \frac{2n\pi t}{\tau} \cos\left(\frac{2n\pi t}{\tau}\right) \right]_0^{\tau/2}$$

$$= \frac{\tau}{n^2\pi^2} (-n\pi \cos n\pi)$$

$$= -\frac{\tau(-1)^n}{n\pi}.$$

So $B_1 = \tau/\pi$, $B_2 = -\tau/(2\pi)$ and $B_3 = \tau/(3\pi)$.

(b) Substituting the coefficients into the given equation and simplifying gives

$$V_3(t) = \frac{\tau}{\pi} \left[\sin\left(\frac{2\pi t}{\tau}\right) - \frac{1}{2} \sin\left(\frac{4\pi t}{\tau}\right) + \frac{1}{3} \sin\left(\frac{6\pi t}{\tau}\right) \right].$$

The graphs of $v(t)$ and $V_3(t)$ on $\left[-\frac{\tau}{2}, \frac{\tau}{2}\right]$ are shown in Figure 21.

The Fourier series for the ramp function $v(t)$ is thus

$$V(t) = -\frac{\tau}{\pi} \sum_{n=1}^{\infty} \frac{(-1)^n}{n} \sin\left(\frac{2n\pi t}{\tau}\right).$$

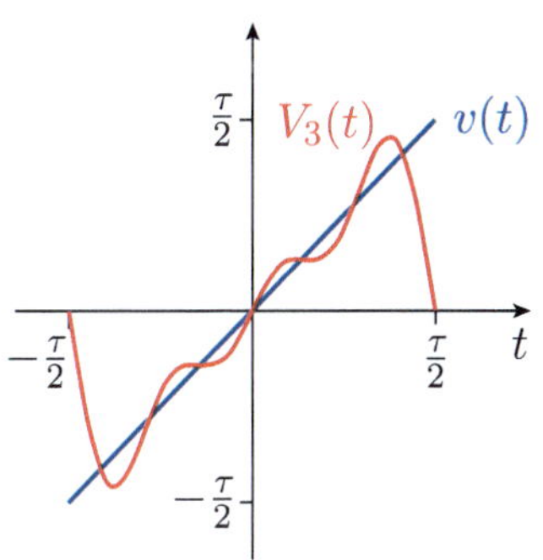

Figure 21   Graphs of $v(t)$ and $V_3(t)$

Our sketch in Figure 21 of the sum of the first three terms gives an approximation to the function, but it is not as good as the corresponding approximation to the sawtooth function that you met in the previous subsection. As in the case of the square-wave function studied in Subsection 2.5, this is due to the discontinuities in the original function. At these points, the Fourier series takes the average value in the middle of the jump. Here that value is 0. In the case of a continuous function, the sizes of the coefficients in the Fourier series generally have an n^2 factor in the denominator and so decrease quite rapidly (such as in equation (30)). By contrast, here and for the square-wave function, where there are discontinuities, the sizes of the coefficients of the Fourier series have only a factor n in the denominator and so the coefficients decrease more slowly. This is a general phenomenon: functions with discontinuities converge more slowly than smooth functions.

Now try to find a Fourier series for an odd function yourself by working through the following exercise.

Exercise 20

The periodic function $f(t)$ with period τ is defined on the interval $\left[-\frac{\tau}{2}, \frac{\tau}{2}\right]$ by

$$f(t) = \begin{cases} -1 & \text{for } -\frac{\tau}{2} < t \le 0, \\ 1 & \text{for } 0 < t < \frac{\tau}{2}. \end{cases}$$

Find the first three non-zero terms of the Fourier series for this function.

4 Fourier series for any periodic function

In the previous section you saw how to find Fourier series for even and odd periodic functions. Unfortunately, not all periodic functions are even or odd. However, the next exercise shows that any function is a sum of an even function and an odd function, so you would expect to be able to approximate functions that are neither even nor odd with a Fourier series involving both sine and cosine terms.

Exercise 21

Consider a general function $f(x)$.

(a) Show that the function $g(x)$ defined by

$$g(x) = \frac{f(x) + f(-x)}{2}$$

is even.

(b) Show that the function $h(x)$ defined by

$$h(x) = \frac{f(x) - f(-x)}{2}$$

is odd.

(c) Show that the function $f(x)$ can be written as the sum

$$f(x) = g(x) + h(x),$$

where $g(x)$ and $h(x)$ are as defined above.

Exercise 21 shows that one way of finding the Fourier series for a general function f is to find Fourier series for the functions g and h as defined in the exercise, and then add them. However, we can find the Fourier series for a general function more directly, as you will see in Subsection 4.1.

The modelling of a real problem may involve a function $f(t)$ that is defined only on some interval. We can choose the interval to be of the form $\left[0, \frac{\tau}{2}\right]$. Then we can extend the definition of the function to the interval $\left[-\frac{\tau}{2}, \frac{\tau}{2}\right]$ by choosing the function to be either even or odd on this interval. From there, we can extend the definition of the function to all the real numbers as a periodic function. You will see how to do this in Subsection 4.2.

4.1 Fourier series for periodic functions

Suppose that you have a periodic function $f(t)$ with period τ and fundamental interval $\left[-\frac{\tau}{2}, \frac{\tau}{2}\right]$. In general, the function will be neither even nor odd. However, it can always be written as a sum of an even function and an odd function, so it should have a Fourier series involving both cosine and sine terms. That is, we can try to represent $f(t)$ as the general Fourier series

$$F(t) = A_0 + \sum_{n=1}^{\infty} A_n \cos\left(\frac{2n\pi t}{\tau}\right) + \sum_{n=1}^{\infty} B_n \sin\left(\frac{2n\pi t}{\tau}\right).$$

As in Subsections 2.4, 3.1 and 3.2, the basic technique is to multiply by cosine or sine terms and integrate over the fundamental interval. Hence we need the following formula that gives the integral when cosine and sine terms are multiplied together.

Trigonometric integrals over the interval $\left[-\frac{\tau}{2}, \frac{\tau}{2}\right]$

For any pair of integers m and n,

$$\int_{-\tau/2}^{\tau/2} \sin\left(\frac{2m\pi t}{\tau}\right) \cos\left(\frac{2n\pi t}{\tau}\right) dt = 0. \tag{38}$$

Using this result, no new terms appear when we form our products, so we arrive at the same formulas as before, which are summarised as follows.

Procedure 1   Fourier series for periodic functions

For a periodic function $f(t)$, with period τ and fundamental interval $\left[-\frac{\tau}{2}, \frac{\tau}{2}\right]$, the Fourier series

$$F(t) = A_0 + \sum_{n=1}^{\infty} A_n \cos\left(\frac{2n\pi t}{\tau}\right) + \sum_{n=1}^{\infty} B_n \sin\left(\frac{2n\pi t}{\tau}\right)$$

is found by using the formulas

$$A_0 = \frac{1}{\tau} \int_{-\tau/2}^{\tau/2} f(t)\, dt,$$

$$A_n = \frac{2}{\tau} \int_{-\tau/2}^{\tau/2} f(t) \cos\left(\frac{2n\pi t}{\tau}\right) dt \quad (n = 1, 2, \ldots),$$

$$B_n = \frac{2}{\tau} \int_{-\tau/2}^{\tau/2} f(t) \sin\left(\frac{2n\pi t}{\tau}\right) dt \quad (n = 1, 2, \ldots).$$

It should be noted that the integrals for determining A_0, A_n and B_n are all over intervals that are symmetric about the origin, which means that the results for integrals of even and odd functions (equations (7) and (8)) will apply. In particular, if f is even then B_n is zero, and if f is odd then A_0 and A_n are zero.

The Fourier coefficients of a given function are unique. So if a function is itself a sum of sine and cosine functions, then it is its own Fourier series. For example, the Fourier series for $\frac{1}{2} \sin 3t$ is $\frac{1}{2} \sin 3t$ – that is, $B_3 = \frac{1}{2}$ and all other Fourier coefficients are zero.

Example 10

The periodic function $f(t)$ is defined by

$$f(t) = e^t \quad \text{for } -1 \leq t \leq 1,$$

on the fundamental interval $[-1, 1]$. Find its Fourier series.

Solution

The function $f(t)$ has the graph shown in Figure 22.

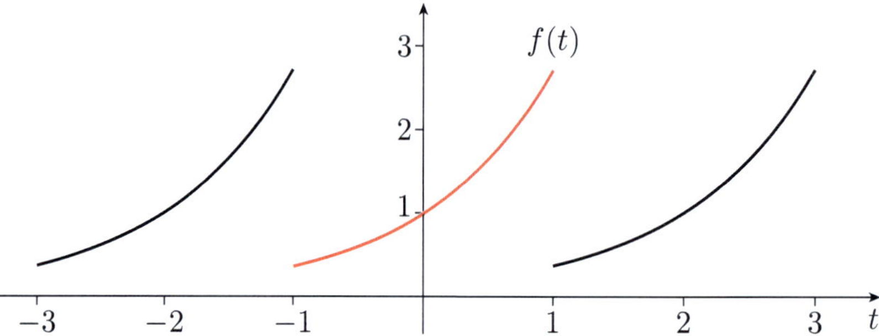

Figure 22   Graph of the function $f(t)$, with the fundamental interval highlighted in red

This function is clearly neither even nor odd. Using Procedure 1, with period $\tau = 2$, we first obtain

$$A_0 = \tfrac{1}{2} \int_{-1}^{1} e^t \, dt = \tfrac{1}{2}(e - e^{-1})$$

and

$$A_n = \int_{-1}^{1} e^t \cos(n\pi t) \, dt.$$

Now use integration by parts (integrating e^t and differentiating $\cos(n\pi t)$) to get

$$A_n = \left[e^t \cos(n\pi t) \right]_{-1}^{1} - \int_{-1}^{1} e^t (-n\pi \sin(n\pi t)) \, dt$$

$$= e \cos n\pi - e^{-1} \cos n\pi + n\pi \int_{-1}^{1} e^t \sin(n\pi t) \, dt. \qquad (39)$$

The integral on the right-hand side of this equation can also be integrated by parts (again integrating e^t and differentiating $\sin(n\pi t)$), so

$$A_n = (e - e^{-1}) \cos n\pi + n\pi \left(\left[e^t \sin(n\pi t) \right]_{-1}^{1} - \int_{-1}^{1} e^t n\pi \cos(n\pi t) \, dt \right).$$

This can be simplified by using $\sin n\pi = 0$ and $\cos n\pi = (-1)^n$ to yield

$$A_n = (e - e^{-1})(-1)^n - n^2 \pi^2 \int_{-1}^{1} e^t \cos(n\pi t) \, dt.$$

The integral on the right-hand side is the integral that we started with for A_n, so we have

$$A_n = (e - e^{-1})(-1)^n - n^2 \pi^2 A_n.$$

Solving this for A_n gives the required coefficient:

$$A_n = \frac{(e - e^{-1})(-1)^n}{1 + n^2 \pi^2}.$$

To find the coefficient B_n, we need to evaluate

$$B_n = \int_{-1}^{1} e^t \sin(n\pi t) \, dt,$$

but this is easier as this integral has already appeared in equation (39) during the computation of A_n. Substituting for A_n in equation (39) gives

$$\frac{(e - e^{-1})(-1)^n}{1 + n^2 \pi^2} = (e - e^{-1})(-1)^n + n\pi B_n.$$

Rearranging this equation and taking out common factors gives

$$B_n = \frac{(e - e^{-1})(-1)^n}{n\pi} \left(\frac{1}{1 + n^2 \pi^2} - 1 \right).$$

Simplifying then gives

$$B_n = -\frac{n\pi(e - e^{-1})(-1)^n}{1 + n^2 \pi^2}.$$

Every coefficient in the Fourier series involves the term $e - e^{-1}$. The series is therefore conveniently written as

$$F(t) = A_0 + \sum_{n=1}^{\infty} A_n \cos\left(\frac{2n\pi t}{\tau}\right) + \sum_{n=1}^{\infty} B_n \sin\left(\frac{2n\pi t}{\tau}\right)$$

$$= (e - e^{-1})\left(\frac{1}{2} + \sum_{n=1}^{\infty} \frac{(-1)^n}{1 + n^2\pi^2}\left(\cos(n\pi t) - n\pi \sin(n\pi t)\right)\right).$$

The graph of the function $f(t)$ and the graph of $F_8(t)$, the sum of the constant and the first eight cosine terms and first eight sine terms in the Fourier series, are shown in Figure 23.

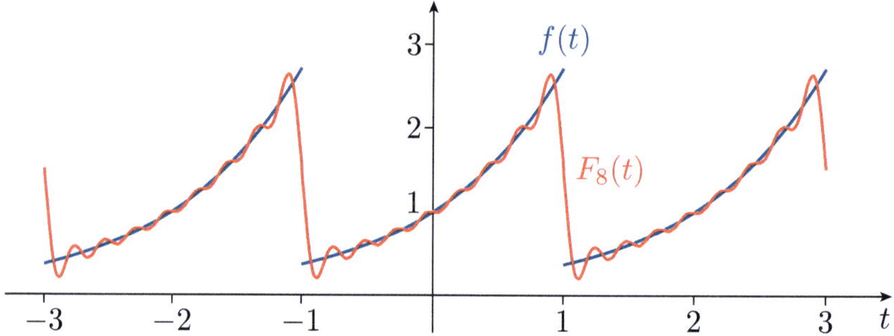

Figure 23   Graph of the function $f(t)$ together with the approximation $F_8(t)$

There is a fairly good approximation to the function, except at the endpoints. At each endpoint, the Fourier series takes the average value, $\frac{1}{2}(e + e^{-1})$, of the function either side of the endpoint.

The following exercises ask you to apply Procedure 1 to find Fourier series.

Exercise 22

Suppose that the periodic function $f(t)$ is defined on the fundamental interval $[-1, 1]$ by

$$f(t) = \begin{cases} 1 & \text{for } -1 \leq t < 0, \\ t & \text{for } 0 \leq t \leq 1. \end{cases}$$

Find the coefficients of its Fourier series.

Exercise 23

(a) Find a fundamental interval for, and hence the period of, the function $b(t)$ defined in the Introduction as

$$b(t) = |\cos t|.$$

(b) State whether $b(t)$ is even, odd, or neither even nor odd, and find the Fourier series for this function.

(*Hint*: Use the trigonometric identity

$$\cos t \cos 2nt = \tfrac{1}{2}\big(\cos(2n-1)t + \cos(2n+1)t\big),$$

which is based on a more general identity given in the Handbook.)

4.2 Functions defined on an interval

So far, you have seen how to calculate the Fourier series for any *periodic* function. However, this is not the whole story. It is also possible to calculate the Fourier series for (almost) *any* function that is defined on a finite interval. This idea will be particularly useful in the next unit.

Suppose that a function $f(t)$ is defined within the finite interval $0 \le t \le T$, such as the curve shown in Figure 24. Furthermore, suppose that we do not care about what happens to the function outside this interval.

Then we can always define another function $f_{\text{ext}}(t)$ to be equal to $f(t)$ on the interval $0 \le t \le T$, and to be periodic with fundamental period T everywhere else. This function is called a **periodic extension** of $f(t)$ and is written as

$$f_{\text{ext}}(t) = f(t) \quad \text{for } 0 \le t < T,$$
$$f_{\text{ext}}(t+T) = f_{\text{ext}}(t).$$

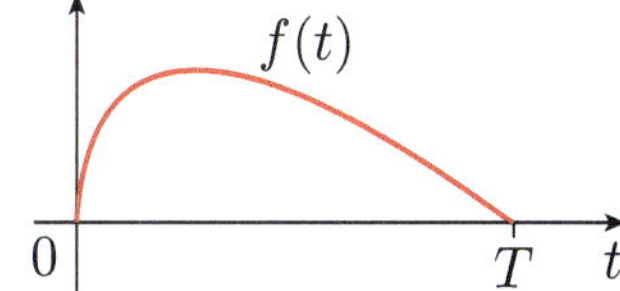

Figure 24   A function $f(t)$ defined on a finite interval $0 \le t \le T$

The graph of $f_{\text{ext}}(t)$ consists of copies of $f(t)$ shifted by T and by all positive and negative integer multiples of T, as shown in Figure 25.

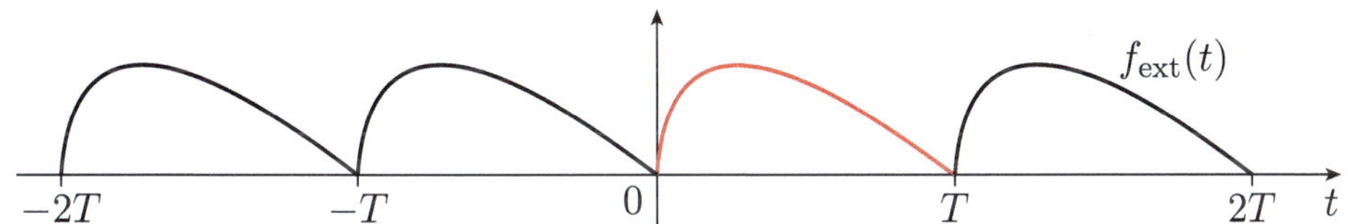

Figure 25   A periodic extension $f_{\text{ext}}(t)$ of $f(t)$

The periodic extension $f_{\text{ext}}(t)$ is a periodic function of fundamental period T, and we can find its Fourier series in the usual way. The resulting Fourier series will be equal to $f_{\text{ext}}(t)$ everywhere, and is equal to $f(t)$ for $0 \le t \le T$. So this Fourier series represents the non-periodic function $f(t)$ inside its domain of definition, $0 \le t \le T$. The periodic extension shown in Figure 25 is neither even nor odd, so the Fourier series contains both sine and cosine terms.

With a little preparation, we can use $f(t)$ to construct periodic functions that are either even or odd, before extending over all t. This is generally a sensible thing to do because the resulting Fourier series will be simpler. The definition of the even extension is straightforward.

Even periodic extension

Consider a function $f(t)$ defined over a finite domain $0 \le t \le T$.

The **even periodic extension** of $f(t)$ is given by

$$f_{\text{even}}(t) = \begin{cases} f(t) & \text{for } 0 \le t < T, \\ f(-t) & \text{for } -T \le t < 0, \end{cases}$$

$$f_{\text{even}}(t + 2T) = f_{\text{even}}(t).$$

An example of this extension is shown in Figure 26.

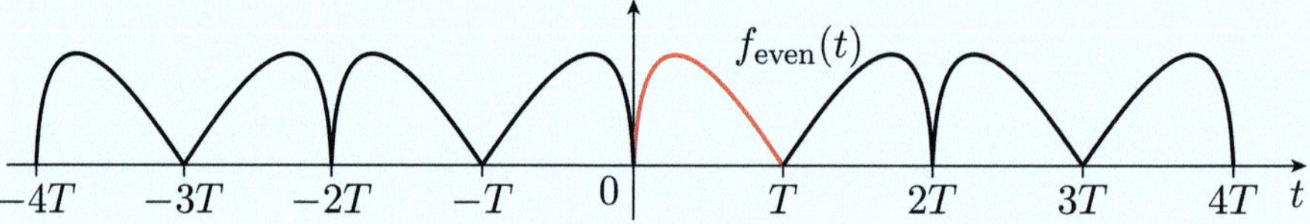

Figure 26   The even periodic extension $f_{\text{even}}(t)$ of $f(t)$

The definition of the odd periodic extension needs a little more care so that the resulting function is both periodic and odd. The reason for this is that any function that is both periodic and odd is zero at the origin and at the endpoints of a fundamental interval centred on the origin.

To show that any odd function $f(t)$ has the value zero at the origin, let a be the value at the origin, that is, $a = f(0)$. Then since f is odd, we have $f(-0) = -f(0)$, thus $a = -a$, that is, $2a = 0$ and so $a = 0$. To show that any odd function with period $2T$ is zero at the right-hand endpoint, let $a = f(T)$. As f is periodic with period $2T$, we must have $f(-T) = f(2T - T) = f(T) = a$. As f is odd, we must have $f(-T) = -f(T)$, so this again leads to the equation $a = -a$, and as before $a = 0$. These results are built into the definition of the odd extension of a function in the following definition.

Odd periodic extension

Consider a function $f(t)$ defined over a finite domain $0 \le t \le T$.

The **odd periodic extension** of $f(t)$ is given by

$$f_{\text{odd}}(t) = \begin{cases} f(t) & \text{for } 0 < t < T, \\ -f(-t) & \text{for } -T < t < 0, \\ 0 & \text{for } t = 0 \text{ or } t = T, \end{cases}$$

$$f_{\text{odd}}(t + 2T) = f_{\text{odd}}(t).$$

An example of this extension is shown in Figure 27.

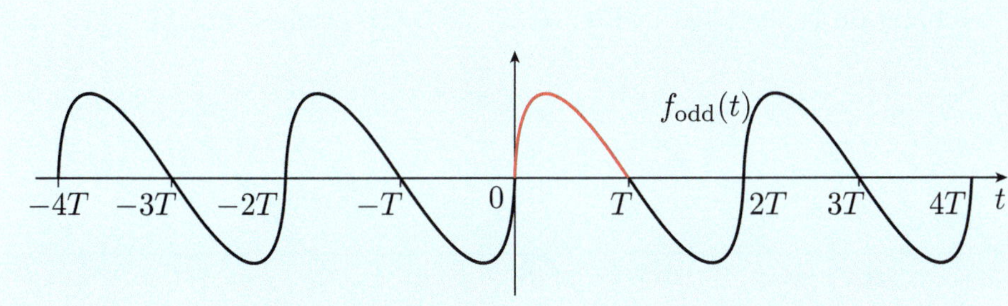

Figure 27   The odd periodic extension $f_{\text{odd}}(t)$ of $f(t)$

In general, both even and odd extensions have fundamental period $\tau = 2T$, but in exceptional cases the even periodic extension may have a smaller fundamental period (see Exercise 25 for an example with fundamental period $\tau = T$).

The next example shows how to use these definitions to define even and odd extensions.

Example 11

Find and sketch the even and odd periodic extensions of the function

$$f(t) = t \quad \text{for } 0 \le t \le 1.$$

Solution

The even periodic extension is given by

$$f_{\text{even}}(t) = \begin{cases} t & \text{for } 0 \le t < 1, \\ -t & \text{for } -1 \le t < 0, \end{cases}$$

$$f_{\text{even}}(t + 2) = f_{\text{even}}(t).$$

This function is sketched in Figure 28, with the original function shown in red.

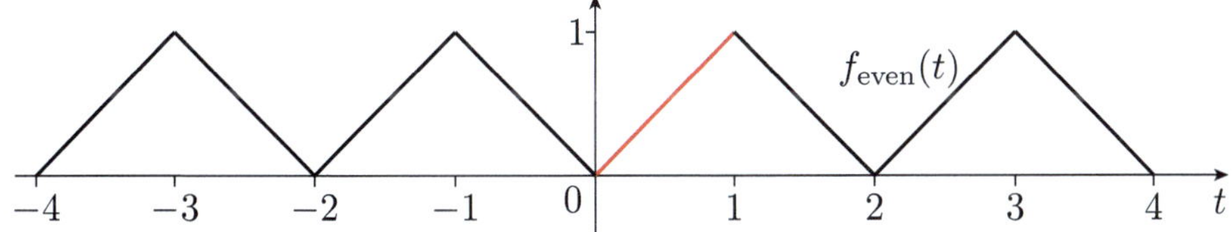

Figure 28   The given function is shown in red and the even periodic extension is shown in black

The odd periodic extension is given by

$$f_{\text{odd}}(t) = \begin{cases} t & \text{for } 0 < t < 1, \\ t & \text{for } -1 < t < 0, \\ 0 & \text{for } t = 0 \text{ or } t = 1 \end{cases}$$

$$f_{\text{odd}}(t + 2) = f_{\text{odd}}(t).$$

This function is sketched in Figure 29, with the original function shown in red.

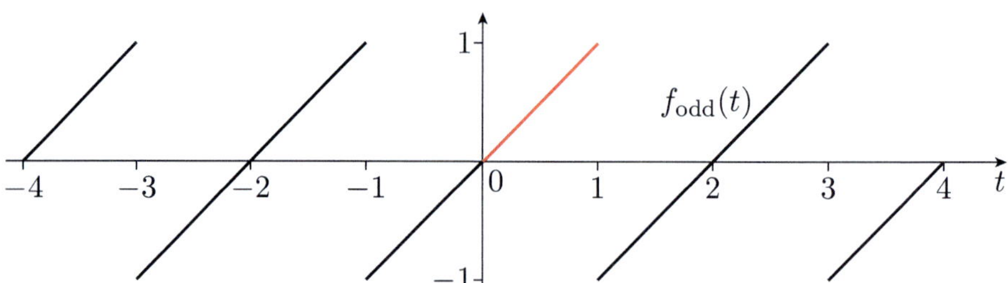

Figure 29 The given function is shown in red and the odd periodic extension is shown in black

In this particular case, both extensions can be expressed in alternative forms. The even periodic extension is

$$f_{\text{even}}(t) = |t| \quad \text{for } -1 \le t < 1,$$
$$f_{\text{even}}(t + 2) = f_{\text{even}}(t),$$

and the odd periodic extension is

$$f_{\text{odd}}(t) = \begin{cases} t & \text{for } -1 < t < 1, \\ 0 & \text{for } t = 1, \end{cases}$$
$$f_{\text{odd}}(t + 2) = f_{\text{odd}}(t).$$

Exercise 24

Consider the function shown in Figure 30 and defined by

$$q(t) = \begin{cases} t & \text{for } 0 \le t \le 1, \\ \frac{3}{2} - \frac{1}{2}t & \text{for } 1 < t \le 3. \end{cases}$$

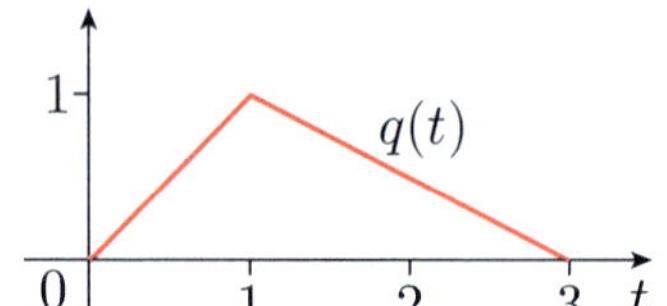

Figure 30

Define the even and odd periodic extensions of $q(t)$, simplifying the formulas if possible. State the fundamental periods, and sketch each extension over a range of three periods.

The following example illustrates how a function defined on a finite interval can be represented by a Fourier series. This result will be used in the next unit.

Example 12

Consider the function defined on the finite interval $0 \le t \le T$ by

$$f(t) = \begin{cases} t/T & \text{for } 0 \le t < T/2, \\ (T - t)/T & \text{for } T/2 \le t \le T, \end{cases}$$

where T is a positive constant. Express $f(t)$ as a Fourier series that involves only sine terms.

Solution

Because we are looking for a Fourier series that involves only sine terms, we need to consider the *odd* periodic extension of $f(t)$, denoted by $f_{\mathrm{odd}}(t)$. This is sketched in Figure 31.

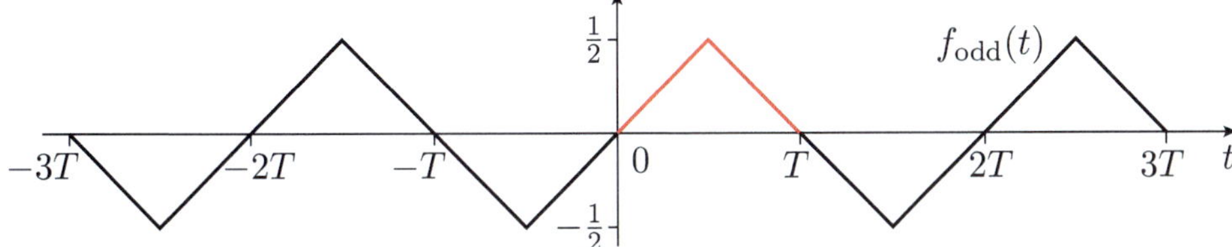

Figure 31   The odd extension of $f(t)$

The function $f_{\mathrm{odd}}(t)$ is odd and has period $\tau = 2T$, so its Fourier series (from equation (32)) takes the form

$$F_{\mathrm{odd}}(t) = \sum_{n=1}^{\infty} B_n \sin\left(\frac{n\pi t}{T}\right),$$

where the Fourier coefficients B_n are given by

$$B_n = \frac{1}{T} \int_{-T}^{T} f_{\mathrm{odd}}(t) \sin\left(\frac{n\pi t}{T}\right) dt.$$

The integrand is even as it is the product of two odd functions, so the integral can be written as twice the integral over positive values. Also, $f_{\mathrm{odd}}(t) = f(t)$ on this interval, so

$$B_n = \frac{2}{T} \int_{0}^{T} f(t) \sin\left(\frac{n\pi t}{T}\right) dt.$$

Using the given piecewise definition of $f(t)$, we obtain

$$B_n = \frac{2}{T^2} \int_{0}^{T/2} t \sin\left(\frac{n\pi t}{T}\right) dt + \frac{2}{T^2} \int_{T/2}^{T} (T - t) \sin\left(\frac{n\pi t}{T}\right) dt.$$

Expanding the second integral gives

$$B_n = \frac{2}{T^2} \int_{0}^{T/2} t \sin\left(\frac{n\pi t}{T}\right) dt + \frac{2}{T} \int_{T/2}^{T} \sin\left(\frac{n\pi t}{T}\right) dt$$
$$- \frac{2}{T^2} \int_{T/2}^{T} t \sin\left(\frac{n\pi t}{T}\right) dt.$$

The first and third integrals are of the form of one of the two useful integrals (equation (17) with $a = n\pi/T$), so we get

$$B_n = \frac{2}{T^2} \left[\frac{T^2}{n^2\pi^2} \left(\sin\left(\frac{n\pi t}{T}\right) - \frac{n\pi t}{T} \cos\left(\frac{n\pi t}{T}\right) \right) \right]_{0}^{T/2}$$
$$+ \frac{2}{T} \left[-\frac{T}{n\pi} \cos\left(\frac{n\pi t}{T}\right) \right]_{T/2}^{T}$$
$$- \frac{2}{T^2} \left[\frac{T^2}{n^2\pi^2} \left(\sin\left(\frac{n\pi t}{T}\right) - \frac{n\pi t}{T} \cos\left(\frac{n\pi t}{T}\right) \right) \right]_{T/2}^{T}.$$

Simplifying,

$$B_n = \frac{2}{n^2\pi^2}\left(\sin\frac{n\pi}{2} - \frac{n\pi}{2}\cos\frac{n\pi}{2}\right) - \frac{2}{n\pi}\left(\cos n\pi - \cos\frac{n\pi}{2}\right)$$
$$- \frac{2}{n^2\pi^2}\left(-n\pi\cos n\pi - \left(\sin\frac{n\pi}{2} - \frac{n\pi}{2}\cos\frac{n\pi}{2}\right)\right).$$

There is a lot of cancellation of terms, and the expression simplifies to

$$B_n = \frac{4}{n^2\pi^2}\sin\left(\frac{n\pi}{2}\right) \quad (n = 1, 2, 3, \ldots).$$

So

$$F_{\text{odd}}(t) = \frac{4}{\pi^2}\sum_{n=1}^{\infty}\frac{1}{n^2}\sin\left(\frac{n\pi}{2}\right)\sin\left(\frac{n\pi t}{T}\right).$$

Since $f(t)$ and $f_{\text{odd}}(t)$ coincide on the interval $0 \le t \le T$, this is the required sine Fourier series $F_{\text{odd}}(t)$ for the odd extension of $f(t)$.

For $n = 1, 2, 3, 4, 5, 6, 7$, the values of $\sin(n\pi/2)$ are $1, 0, -1, 0, 1, 0, -1$, so the first few terms in the Fourier series are

$$F_{\text{odd}}(t) = \frac{4}{\pi^2}\left(\sin(\pi t) - \frac{1}{3^2}\sin(3\pi t) + \frac{1}{5^2}\sin(5\pi t) - \frac{1}{7^2}\sin(7\pi t) + \cdots\right).$$

Exercise 25

Consider the same function $f(t)$ as that discussed in Example 12. Within its domain of definition, $0 \le t \le T$, represent this function by a Fourier series that involves only constant and cosine functions.

To represent the original function $f(t)$ in Example 12 by a Fourier series, we can use the odd periodic extension, obtaining a series that contains only sine terms (as in Example 12), or we can use the even periodic extension, obtaining a series that contains only constant and cosine terms (as in Exercise 25).

Sometimes one choice is better than the other. In general, if we want to approximate a function by a truncated Fourier series, it is better to use a periodic extension that is continuous, rather than discontinuous. This is because, as pointed out earlier, the Fourier series for a continuous function converges more rapidly than that for a discontinuous function. So for the function discussed in Example 11 we would use the even periodic extension to obtain the Fourier series.

However, in the next unit we will use Fourier series to solve partial differential equations, and in that case our choice of an even or odd periodic extension is generally dictated by other factors, namely the boundary conditions.

Exercise 25 is an exceptional case in which the even periodic extension has fundamental period $\tau = T$ rather than $\tau = 2T$. This simplifies the calculations because we need integrals only over the range from 0 to $T/2$. If we were to treat the function in Exercise 25 as having period $2T$, then the usual formula would eventually give the *same* Fourier series, although the calculations would be longer. In general, making the mistake of using a non-fundamental period rather than the fundamental period will always give the same final Fourier series, but at the expense of more labour.

Exercise 26

Consider the function

$$f(t) = \begin{cases} 1 & \text{for } 0 \le t < \frac{\pi}{2}, \\ -1 & \text{for } \frac{\pi}{2} \le t < \pi. \end{cases}$$

(a) Define the even periodic extension, simplifying your answer as much possible. Sketch this function over $-3\pi \le t \le 3\pi$, and state its fundamental period.

(b) Find the Fourier series for the even periodic extension.

(c) Define the odd periodic extension. Sketch this function over $-3\pi \le t \le 3\pi$, and state its fundamental period.

Learning outcomes

After studying this unit, you should be able to:

- understand the terms frequency, period and fundamental interval, and obtain them for a periodic function

- understand the terms even and odd as applied to functions, and test a function to see if it is either

- find the Fourier series for a periodic function

- compare the graph of a function with the graph of a sum of terms in the Fourier series, and comment on the closeness of the approximation to the function

- understand how to modify a function defined on an interval to give an even or odd periodic extension.

Solutions to exercises

Solution to Exercise 1

We have

$$
\begin{aligned}
G(t + 2\pi) &= \frac{1}{2} + \frac{4}{\pi^2}\left(\cos(t + 2\pi) + \frac{1}{9}\cos 3(t + 2\pi)\right.\\
&\qquad\left. + \frac{1}{25}\cos 5(t + 2\pi) + \frac{1}{49}\cos 7(t + 2\pi) + \cdots\right)\\
&= \frac{1}{2} + \frac{4}{\pi^2}\left(\cos(t + 2\pi) + \frac{1}{9}\cos(3t + 6\pi)\right.\\
&\qquad\left. + \frac{1}{25}\cos(5t + 10\pi) + \frac{1}{49}\cos(7t + 14\pi) + \cdots\right)\\
&= \frac{1}{2} + \frac{4}{\pi^2}\left(\cos t + \frac{1}{9}\cos 3t + \frac{1}{25}\cos 5t + \frac{1}{49}\cos 7t + \cdots\right)\\
&= G(t).
\end{aligned}
$$

Note that $G(t)$ would also be unchanged by adding any integer multiple of 2π to its argument. These results hold because adding 2π any number of times to a cosine argument does not change the value of the cosine.

Solution to Exercise 2

The angular frequencies are 1, 3, 5, 7, …,

and the periods are 2π, $\frac{2\pi}{3}$, $\frac{2\pi}{5}$, $\frac{2\pi}{7}$, ….

Solution to Exercise 3

The angular frequencies of the component functions are π, $\frac{3\pi}{2}$, 2π,

and the corresponding periods are 2, $\frac{4}{3}$, 1.

The least common multiple of these periods is 4, so the period of the function $f(t)$ is $\tau = 4$.

Solution to Exercise 4

The two graphs are as follows.

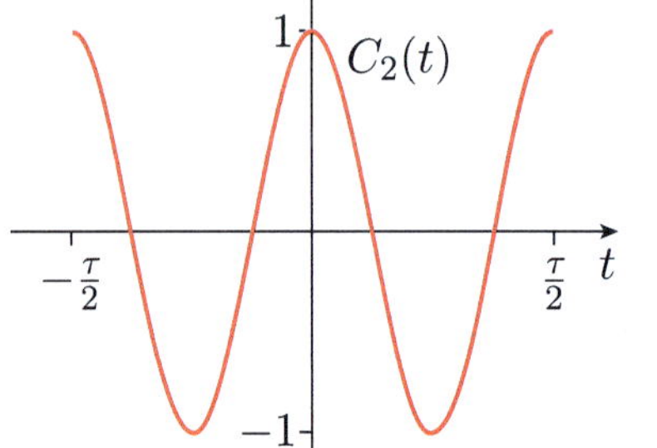

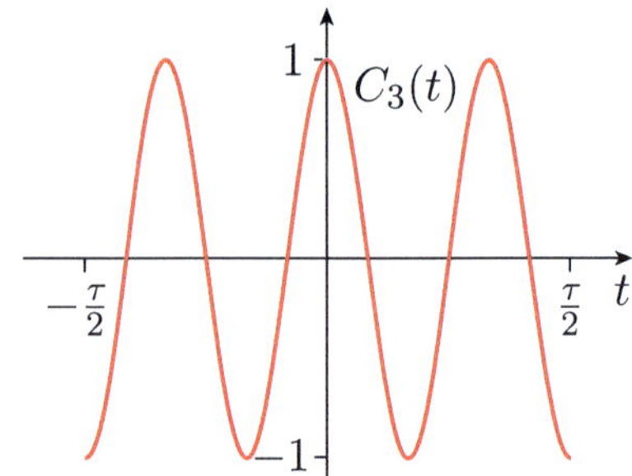

When $n = 0$, formula (6) gives

$$C_0(t) = \cos\left(\frac{2 \times 0\pi t}{\tau}\right) = \cos 0 = 1.$$

This is a constant function, which is periodic.

Solution to Exercise 5

Since $g(-t) = (-t)^3 = -t^3 = -g(t)$ for al t, the function is odd.

Solution to Exercise 6

Since $f(t)$ and $g(t)$ are both odd functions,

$$f(-t) = -f(t), \quad g(-t) = -g(t).$$

Hence

$$k(-t) = f(-t) + g(-t) = -f(t) + (-g(t)) = -(f(t) + g(t)) = -k(t),$$

so $k(t)$ is an odd function.

Solution to Exercise 7

(a) The graph of the function $C_0(t)$ is shown below. As the graph is symmetrical about the vertical axis, the function must be even.

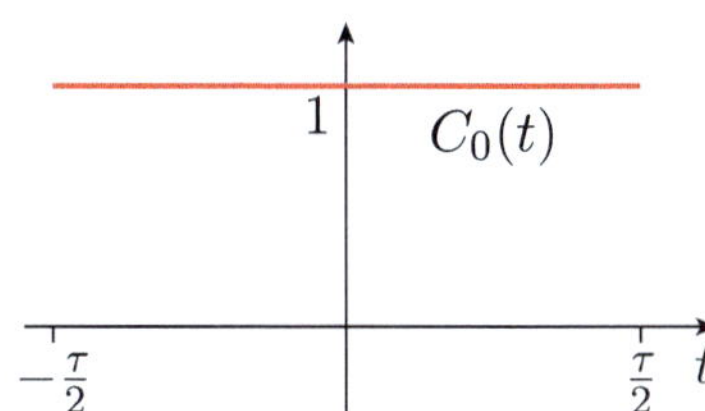

(b) Here $h(-t) = (-t)^2 + (-t)^3 = t^2 - t^3$. In general, this is equal to neither $h(t)$ nor $-h(t)$, so the function is neither even nor odd. This can also be seen by the asymmetrical nature of the curve in the diagram below.

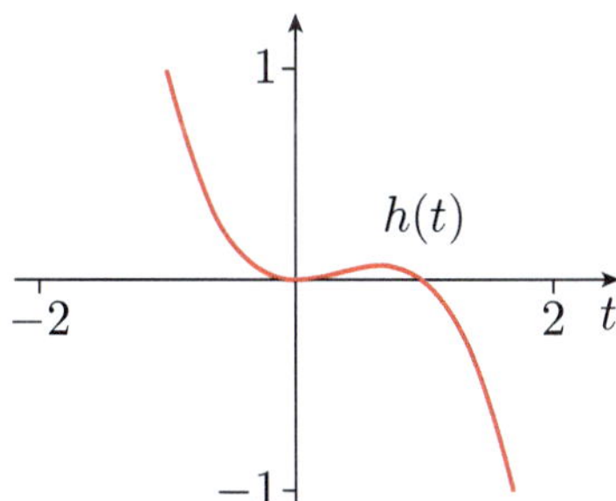

Solution to Exercise 8

We have

$$h(t) = f(t)\, g(t) = (t^3 + 2t^5)(3t^2 - t^4)$$
$$= 3t^5 - t^7 + 6t^7 - 2t^9$$
$$= 3t^5 + 5t^7 - 2t^9.$$

All the powers of t are odd, so $h(t)$ is an odd function.

Alternatively, since $f(t)$ and $g(t)$ are odd and even functions, respectively, we know by definition that

$$f(-t) = -f(t), \quad g(-t) = g(t).$$

Hence

$$h(-t) = f(-t)\, g(-t) = -f(t)\, g(t) = -h(t),$$

so $h(t)$ is an odd function.

Solution to Exercise 9

From the graph, the values of the function repeat after an interval of length 2π. Hence the period is $\tau = 2\pi$. The angular frequency ω satisfies $\tau = 2\pi/\omega$, so here $\omega = 1$.

Solution to Exercise 10

For $n = 1, 2, 3, \ldots$,

$$\int_{-\pi}^{\pi} \cos nt\, dt = \left[\frac{\sin nt}{n} \right]_{-\pi}^{\pi} = 0,$$

since $\sin n\pi = 0$ and $\sin(-n\pi) = 0$.

Solution to Exercise 11

Applying equation (13) gives

$$A_0 = \frac{1}{2\pi} \int_{-\pi}^{\pi} h(t)\, dt.$$

Using the hint, this simplifies to

$$A_0 = \frac{1}{2\pi} \int_{-\pi/2}^{\pi/2} 1\, dt,$$

where we have used the fact that $h(t) = 1$ for $-\pi/2 \le t \le \pi/2$. Performing the integration gives

$$A_0 = \frac{1}{2\pi} [t]_{-\pi/2}^{\pi/2} = \tfrac{1}{2}.$$

Solution to Exercise 12

(a) As A_0 is a constant, it can be taken outside the integral to obtain

$$\int_{-\pi}^{\pi} A_0 \cos t \, dt = A_0 \int_{-\pi}^{\pi} \cos t \, dt = A_0 \left[\sin t \right]_{-\pi}^{\pi} = 0.$$

(b) Using the given trigonometric identity with $\alpha = 2t$ and $\beta = t$, we have

$$\int_{-\pi}^{\pi} \cos 2t \cos t \, dt = \int_{-\pi}^{\pi} \left(\tfrac{1}{2} \cos 3t + \tfrac{1}{2} \cos t \right) dt$$
$$= \left[\tfrac{1}{6} \sin 3t + \tfrac{1}{2} \sin t \right]_{-\pi}^{\pi}$$
$$= 0.$$

(c) Again using the trigonometric identity, this time with $\alpha = nt$ and $\beta = t$, we obtain

$$\int_{-\pi}^{\pi} \cos nt \cos t \, dt = \int_{-\pi}^{\pi} \left(\tfrac{1}{2} \cos(n+1)t + \tfrac{1}{2} \cos(n-1)t \right) dt$$
$$= \left[\frac{\sin(n+1)t}{2(n+1)} + \frac{\sin(n-1)t}{2(n-1)} \right]_{-\pi}^{\pi} \quad (\text{as } n > 1)$$
$$= 0.$$

(d) Rearranging the given trigonometric identity to make $\cos^2 \alpha$ the subject, and using $\alpha = t$, we have

$$\int_{-\pi}^{\pi} \cos^2 t \, dt = \int_{-\pi}^{\pi} \tfrac{1}{2}(\cos 2t + 1) \, dt$$
$$= \left[\tfrac{1}{4} \sin 2t + \tfrac{1}{2}t \right]_{-\pi}^{\pi}$$
$$= \pi.$$

Solution to Exercise 13

Substitute the definition of $h(t)$ from Exercise 11 into equation (16) to obtain

$$A_1 = \frac{1}{\pi} \int_{-\pi}^{\pi} h(t) \cos t \, dt.$$

Now use the definition of $h(t)$ to get (as $h(t) = 0$ outside the interval $[-\frac{\pi}{2}, \frac{\pi}{2}]$)

$$A_1 = \frac{1}{\pi} \int_{-\pi/2}^{\pi/2} 1 \times \cos t \, dt.$$

Performing the integration gives

$$A_1 = \frac{1}{\pi} \left[\sin t \right]_{-\pi/2}^{\pi/2} = \frac{1}{\pi}(1 - (-1)) = \frac{2}{\pi}.$$

Solution to Exercise 14

We could obtain a general formula for A_n, as in Example 6, and then substitute $n = 2$ and $n = 3$ into that. Alternatively, put $n = 2$ into equation (22) and obtain

$$A_2 = \frac{1}{\pi} \int_{-\pi}^{\pi} h(t) \cos 2t \, dt.$$

Now use the definition of $h(t)$, which is zero outside $-\pi/2 \le t \le \pi/2$ and equal to 1 on this interval, to yield

$$A_2 = \frac{1}{\pi} \int_{-\pi/2}^{\pi/2} 1 \times \cos 2t \, dt.$$

Performing the integration gives

$$A_2 = \frac{1}{\pi} \left[\tfrac{1}{2} \sin 2t \right]_{-\pi/2}^{\pi/2} = \frac{1}{\pi} \left(\tfrac{1}{2} \sin \pi - \tfrac{1}{2} \sin(-\pi) \right) = 0.$$

Substituting $n = 3$ into equation (22) gives

$$A_3 = \frac{1}{\pi} \int_{-\pi}^{\pi} h(t) \cos 3t \, dt.$$

Now use the definition of $h(t)$ again to yield

$$A_3 = \frac{1}{\pi} \int_{-\pi/2}^{\pi/2} 1 \times \cos 3t \, dt.$$

Performing the integration gives

$$A_3 = \frac{1}{\pi} \left[\tfrac{1}{3} \sin 3t \right]_{-\pi/2}^{\pi/2} = \frac{1}{3\pi} \left(\sin \tfrac{3\pi}{2} - \sin\left(-\tfrac{3\pi}{2}\right) \right)$$
$$= \frac{1}{3\pi} \left((-1) - 1 \right) = -\frac{2}{3\pi}.$$

Solution to Exercise 15

The constant term does not fit in the general pattern, so we can leave this term outside the summation. All of the other terms have a factor 2 in the numerator and a factor π in the denominator, so we can factorise the expression as

$$H(t) = \frac{1}{2} + \frac{2}{\pi} \left(\cos t - \frac{1}{3} \cos 3t + \frac{1}{5} \cos 5t - \frac{1}{7} \cos 7t + \cdots \right).$$

The sum in the brackets is a sum over all odd numbers. As $n = 1, 2, 3, \ldots$, the expression $2n - 1$ evaluates to $1, 3, 5, \ldots$, so this is the expression to be used in the angular frequency and the denominator. The terms also change sign, so we need a factor $(-1)^n$ in the coefficients. The first sign is positive, and $(-1)^n$ is negative for $n = 1$, so we need an extra minus sign (either by writing this as $(-1)^{n+1}$ or by taking a minus sign out of the bracket as below).

Putting all this together gives the closed form as

$$H(t) = \frac{1}{2} - \frac{2}{\pi} \sum_{n=1}^{\infty} \frac{(-1)^n}{2n - 1} \cos(2n - 1)t.$$

Solution to Exercise 16

We have

$$A_0 = \frac{1}{2\pi} \int_{-\pi}^{\pi} f(t)\, dt = \frac{1}{2\pi} \int_{-\pi}^{\pi} t^2\, dt = \frac{1}{2\pi} \left[\tfrac{1}{3} t^3\right]_{-\pi}^{\pi} = \tfrac{1}{3}\pi^2,$$

and for $n = 1, 2, 3, \ldots$, we use the hint to obtain

$$\begin{aligned}
A_n &= \frac{1}{\pi} \int_{-\pi}^{\pi} t^2 \cos nt\, dt \\
&= \frac{1}{\pi} \left[\frac{1}{n^3} \left((n^2 t^2 - 2) \sin nt + 2nt \cos nt \right) \right]_{-\pi}^{\pi} \\
&= \frac{1}{\pi} \times \frac{1}{n^3} \left(2n\pi(-1)^n - (-2n\pi(-1)^n) \right),
\end{aligned}$$

where we have used $\sin n\pi = 0$ and $\cos n\pi = (-1)^n$ as n is an integer. This simplifies to

$$A_n = \frac{4(-1)^n}{n^2}.$$

The Fourier series is therefore

$$\begin{aligned}
F(t) &= \frac{\pi^2}{3} + 4 \sum_{n=1}^{\infty} \frac{(-1)^n}{n^2} \cos nt \\
&= \frac{\pi^2}{3} - 4 \cos t + \cos 2t - \frac{4}{9} \cos 3t + \frac{1}{4} \cos 4t + \cdots.
\end{aligned}$$

(The graph of $F(t)$ matches the expected graph, which consists of a parabola on the fundamental interval repeated indefinitely to both the left and the right – see below.)

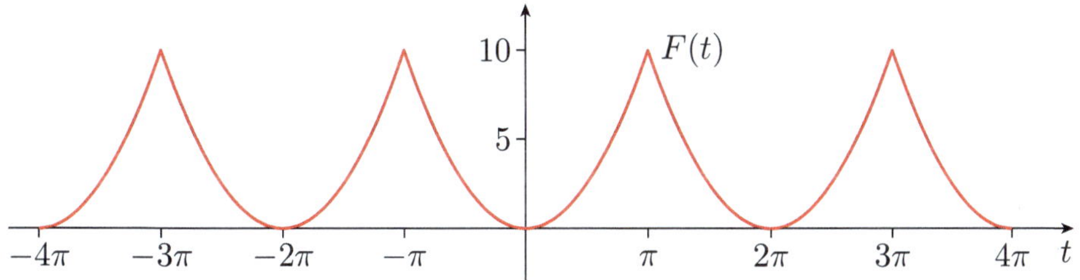

Solution to Exercise 17

The function $w(t)$ is even, and $w(t) = t$ on the interval $[0, \pi]$. Therefore

$$A_0 = \frac{1}{2\pi} \int_{-\pi}^{\pi} w(t)\, dt = \frac{1}{\pi} \int_{0}^{\pi} t\, dt = \frac{1}{\pi} \left[\tfrac{1}{2} t^2\right]_{0}^{\pi} = \frac{\pi}{2}.$$

For the other coefficients we must evaluate

$$A_n = \frac{1}{\pi} \int_{-\pi}^{\pi} w(t) \cos nt\, dt.$$

As $w(t)$ is even and equal to t for t positive, this simplifies to

$$A_n = \frac{2}{\pi} \int_{0}^{\pi} t \cos nt\, dt.$$

This is one of the useful integrals (equation (18) with $a = n$), so

$$A_n = \frac{2}{\pi} \times \frac{1}{n^2} \left[\cos nt + nt \sin nt \right]_0^\pi.$$

Evaluating this expression (and using $\sin n\pi = 0$ and $\cos n\pi = (-1)^n$), we get

$$A_n = \frac{2}{n^2 \pi} \left((-1)^n - 1 \right).$$

Thus the Fourier series is

$$W(t) = \frac{\pi}{2} + \frac{2}{\pi} \sum_{n=1}^\infty \frac{(-1)^n - 1}{n^2} \cos nt.$$

So the requested approximation is

$$W_5(t) = \frac{\pi}{2} - \frac{4}{\pi} \cos t - \frac{4}{9\pi} \cos 3t - \frac{4}{25\pi} \cos 5t.$$

Solution to Exercise 18

The easiest way to proceed is to draw a graph of the given function on the fundamental interval, such as in the figure in the margin. The pointwise convergence theorem gives

$$F(-1) = \frac{f(-1^+) + f(-1^-)}{2}.$$

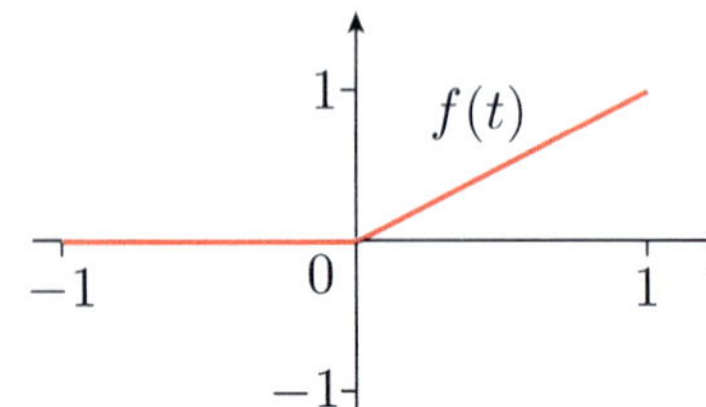

Using the graph as a guide, we see that the given function is discontinuous at the point $t = -1$ and that approaching the point from the right gives $f(-1^+) = 0$. Approaching $t = -1$ from the left is outside the fundamental interval, so we add one period to see that this has the same value as approaching $t = 1$ from the left, that is, $f(-1^-) = f(1^-) = 1$. So

$$F(-1) = \frac{0 + 1}{2} = \frac{1}{2}.$$

The given function $f(t)$ is continuous at $t = 0$, so $F(t) = f(t)$ at $t = 0$, that is,

$$F(0) = f(0) = 0.$$

(You may observe that the slope of the curve changes abruptly at $t = 0$, so the derivative of $f(t)$ is discontinuous at $t = 0$. This property is independent of the fact that $f(t)$ is continuous at $t = 0$.)

Solution to Exercise 19

Using equation (28) gives

$$A_0 = \frac{1}{\tau} \int_{-\tau/2}^{\tau/2} w(t) \, dt.$$

As $w(t)$ is even, this integral can be simplified and evaluated as

$$A_0 = \frac{2}{\tau} \int_0^{\tau/2} t \, dt = \frac{2}{\tau} \left[\tfrac{1}{2} t^2 \right]_0^{\tau/2} = \tfrac{1}{4}\tau.$$

Similarly, we can compute A_n by using equation (29):

$$A_n = \frac{2}{\tau} \int_{-\tau/2}^{\tau/2} w(t) \cos\left(\frac{2n\pi t}{\tau}\right) dt.$$

Proceeding as before, we can simplify the integral by using the fact that the integrand $w(t)\cos(2n\pi t/\tau)$ is even, to get

$$A_n = \frac{4}{\tau} \int_0^{\tau/2} w(t) \cos\left(\frac{2n\pi t}{\tau}\right) dt$$

$$= \frac{4}{\tau} \int_0^{\tau/2} t \cos\left(\frac{2n\pi t}{\tau}\right) dt.$$

This integral is again of the form of one of the two useful integrals (equation (18) with $a = 2n\pi/\tau$), so

$$A_n = \frac{4}{\tau} \times \frac{\tau^2}{4n^2\pi^2} \left[\cos\left(\frac{2n\pi t}{\tau}\right) + \frac{2n\pi t}{\tau}\sin\left(\frac{2n\pi t}{\tau}\right)\right]_0^{\tau/2}.$$

Evaluating this expression (and using $\sin n\pi = 0$ and $\cos n\pi = (-1)^n$) gives

$$A_n = \frac{\tau}{n^2\pi^2}\left(\left(\cos n\pi + n\pi \sin n\pi\right) - (1 - 0)\right)$$

$$= \frac{\tau((-1)^n - 1)}{n^2\pi^2}.$$

(So we have

$$A_0 = \frac{\tau}{4}, \quad A_1 = -\frac{2\tau}{\pi^2}, \quad A_2 = 0, \quad A_3 = -\frac{2\tau}{9\pi^2}, \quad A_4 = 0,$$

$$A_5 = -\frac{2\tau}{25\pi^2}, \quad A_6 = 0, \quad A_7 = -\frac{2\tau}{49\pi^2}, \quad A_8 = 0, \quad \dots .)$$

Solution to Exercise 20

The function is odd, so its Fourier series involves only sine terms:

$$F(t) = \sum_{n=1}^{\infty} B_n \sin\left(\frac{2n\pi t}{\tau}\right),$$

where the coefficients are given by

$$B_n = \frac{2}{\tau} \int_{-\tau/2}^{\tau/2} f(t) \sin\left(\frac{2n\pi t}{\tau}\right) dt.$$

As $f(t)$ is odd, the integrand is even (as the product of two odd functions). In addition, $f(t)$ is equal to 1 for positive t, so the integral simplifies to

$$B_n = \frac{4}{\tau} \int_0^{\tau/2} \sin\left(\frac{2n\pi t}{\tau}\right) dt.$$

Evaluating the integral gives

$$B_n = \frac{4}{\tau}\left[-\frac{\tau}{2n\pi}\cos\left(\frac{2n\pi t}{\tau}\right)\right]_0^{\tau/2} = \frac{2\left(-\cos n\pi - (-1)\right)}{n\pi},$$

$$= \frac{2(1 - (-1)^n)}{n\pi}.$$

Evaluating each coefficient in turn yields

$$\frac{4}{\pi}, \; 0, \; \frac{4}{3\pi}, \; 0, \; \frac{4}{5\pi}, \; 0, \; \ldots,$$

so the first three non-zero terms of the Fourier series for $f(t)$ are

$$F(t) = \frac{4}{\pi}\left(\sin\left(\frac{2\pi t}{\tau}\right) + \frac{1}{3}\sin\left(\frac{6\pi t}{\tau}\right) + \frac{1}{5}\sin\left(\frac{10\pi t}{\tau}\right) + \cdots\right).$$

(Note that since $B_n = 0$ for even n, the Fourier series could be written as a sum over odd terms,

$$F(t) = \frac{4}{\pi}\sum_{n=1}^{\infty}\frac{1}{2n-1}\sin\left(\frac{2(2n-1)\pi t}{\tau}\right),$$

but this was not required in this exercise.)

Solution to Exercise 21

(a) Using the definition of $g(x)$, we have

$$g(-x) = \frac{f(-x) + f(-(-x))}{2} = \frac{f(-x) + f(x)}{2} = g(x).$$

Hence $g(x)$ is an even function.

(b) Using the definition of $h(x)$, we have

$$h(-x) = \frac{f(-x) - f(-(-x))}{2} = \frac{f(-x) - f(x)}{2} = -h(x).$$

Hence $h(x)$ is an odd function.

(c) Using the definitions of $g(x)$ and $h(x)$, we have

$$g(x) + h(x) = \frac{f(x) + f(-x)}{2} + \frac{f(x) - f(-x)}{2} = f(x).$$

Solution to Exercise 22

Using Procedure 1, we obtain

$$A_0 = \tfrac{1}{2}\int_{-1}^{0} 1\,dt + \tfrac{1}{2}\int_{0}^{1} t\,dt$$

$$= \tfrac{1}{2}\left[t\right]_{-1}^{0} + \tfrac{1}{2}\left[\tfrac{1}{2}t^2\right]_{0}^{1} = \tfrac{1}{2} + \tfrac{1}{4} = \tfrac{3}{4},$$

$$A_n = \int_{-1}^{0} \cos(n\pi t)\,dt + \int_{0}^{1} t\cos(n\pi t)\,dt$$

Using equation (18) with $a = n\pi$.

$$= \frac{1}{n\pi}\left[\sin(n\pi t)\right]_{-1}^{0} + \left[\frac{1}{n^2\pi^2}\left(\cos(n\pi t) + n\pi t\sin(n\pi t)\right)\right]_{0}^{1}$$

$$= \frac{1}{n^2\pi^2}\left((-1)^n - 1\right),$$

$$B_n = \int_{-1}^{0} \sin(n\pi t)\, dt + \int_{0}^{1} t \sin(n\pi t)\, dt$$

$$= -\frac{1}{n\pi}\Big[\cos(n\pi t)\Big]_{-1}^{0} + \left[\frac{1}{n^2\pi^2}\big(\sin(n\pi t) - n\pi t \cos(n\pi t)\big)\right]_{0}^{1} \qquad \text{Using equation (17) with } a = n\pi.$$

$$= -\frac{1}{n\pi}(1 - \cos n\pi) - \frac{1}{n^2\pi^2}\big(n\pi \cos n\pi\big)$$

$$= -\frac{1}{n\pi} + \frac{\cos n\pi}{n\pi} - \frac{\cos n\pi}{n\pi}$$

$$= -\frac{1}{n\pi}.$$

Solution to Exercise 23

(a) It is clear from the graph of $b(t)$ (Figure 2) that $\left[-\frac{\pi}{2}, \frac{\pi}{2}\right]$ is a fundamental interval, so $b(t)$ has period π.

(b) The function $b(t)$ is even, so its Fourier series involves only the constant and cosine terms, hence

$$B(t) = A_0 + \sum_{n=1}^{\infty} A_n \cos 2nt.$$

Starting with the constant term, we need to calculate

$$A_0 = \frac{1}{\pi} \int_{-\pi/2}^{\pi/2} \cos t \, dt.$$

Evaluating the integral gives

$$A_0 = \frac{1}{\pi}\big[\sin t\big]_{-\pi/2}^{\pi/2} = \frac{1}{\pi}\big(1 - (-1)\big) = \frac{2}{\pi}.$$

Similarly, the coefficient A_n can be calculated from

$$A_n = \frac{2}{\pi} \int_{-\pi/2}^{\pi/2} \cos t \cos 2nt \, dt.$$

As the integrand is even, this integral simplifies to

$$A_n = \frac{4}{\pi} \int_{0}^{\pi/2} \cos t \cos 2nt \, dt.$$

Using the hint, we write this as

$$A_n = \frac{4}{\pi} \int_{0}^{\pi/2} \tfrac{1}{2}\big(\cos(2n-1)t + \cos(2n+1)t\big) \, dt.$$

Evaluating the integral gives

$$A_n = \frac{2}{\pi}\left[\frac{\sin(2n-1)t}{2n-1} + \frac{\sin(2n+1)t}{2n+1}\right]_{0}^{\pi/2}$$

$$= \frac{2}{\pi}\left(\frac{\sin(n\pi - \frac{\pi}{2})}{2n-1} + \frac{\sin(n\pi + \frac{\pi}{2})}{2n+1}\right).$$

Now use the addition formulas for sine to get

$$A_n = \frac{2}{\pi} \left(\frac{\sin n\pi \cos \frac{\pi}{2} - \cos n\pi \sin \frac{\pi}{2}}{2n-1} + \frac{\sin n\pi \cos \frac{\pi}{2} + \cos n\pi \sin \frac{\pi}{2}}{2n+1} \right).$$

Using $\sin n\pi = 0$, $\cos n\pi = (-1)^n$ and $\sin \frac{\pi}{2} = 1$, this simplifies to

$$\begin{aligned}
A_n &= \frac{2}{\pi} \left(\frac{-(-1)^n}{2n-1} + \frac{(-1)^n}{2n+1} \right) \\
&= \frac{2(-1)^n}{\pi} \left(-\frac{1}{2n-1} + \frac{1}{2n+1} \right) \\
&= \frac{2(-1)^n}{\pi} \times \frac{-(2n+1) + (2n-1)}{(2n-1)(2n+1)} \\
&= -\frac{4(-1)^n}{\pi(4n^2-1)}.
\end{aligned}$$

So the Fourier series is

$$\begin{aligned}
B(t) &= \frac{2}{\pi} - \frac{4}{\pi} \sum_{n=1}^{\infty} \frac{(-1)^n}{4n^2-1} \cos 2nt \\
&= \frac{2}{\pi} \left(1 + \frac{2}{3} \cos 2t - \frac{2}{15} \cos 4t + \frac{2}{35} \cos 6t - \frac{2}{63} \cos 8t + \cdots \right).
\end{aligned}$$

Solution to Exercise 24

Using the definition of the even periodic extension, we have

$$q_{\text{even}}(t) = \begin{cases} t & \text{for } 0 \le t \le 1, \\ \frac{3}{2} - \frac{1}{2}t & \text{for } 1 < t \le 3, \\ -t & \text{for } -1 \le t < 0, \\ \frac{3}{2} + \frac{1}{2}t & \text{for } -3 < t < -1, \end{cases}$$

$$q_{\text{even}}(t+6) = q_{\text{even}}(t).$$

This function has fundamental period $\tau = 6$, and its formula cannot be made much simpler. Its graph is sketched below.

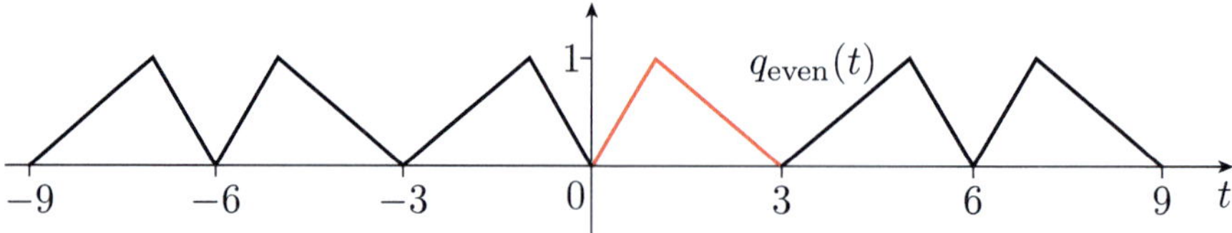

The odd periodic extension is given by

$$q_{\text{odd}}(t) = \begin{cases} t & \text{for } 0 \le t \le 1, \\ \frac{3}{2} - \frac{1}{2}t & \text{for } 1 < t \le 3, \\ t & \text{for } -1 \le t < 0, \\ -\frac{3}{2} - \frac{1}{2}t & \text{for } -3 < t < -1, \end{cases}$$

$$q_{\text{odd}}(t+6) = q_{\text{odd}}(t).$$

This function has fundamental period $\tau = 6$, and its graph is sketched below.

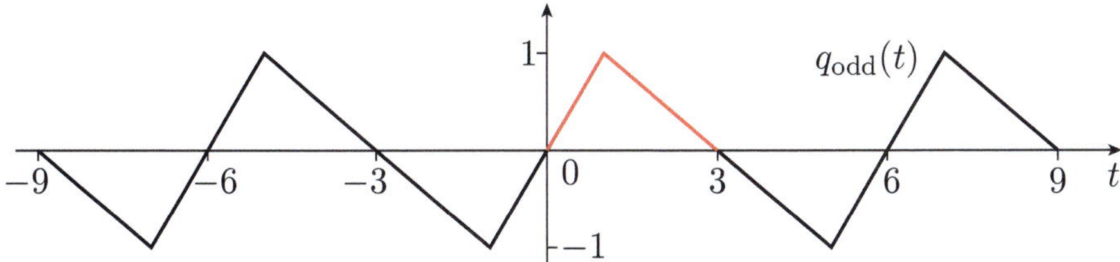

By examining this graph, we see that the formula can be simplified to

$$q_{\text{odd}}(t) = \begin{cases} t & \text{for } -1 \leq t < 1, \\ \frac{3}{2} - \frac{1}{2}t & \text{for } 1 \leq t < 5, \end{cases}$$

$$q_{\text{odd}}(t+6) = q_{\text{odd}}(t).$$

Solution to Exercise 25

Because we are looking for a Fourier series that involves only constant and cosine terms, we need to consider the *even* periodic extension of $f(t)$, denoted by $f_{\text{even}}(t)$. This is sketched in the figure below.

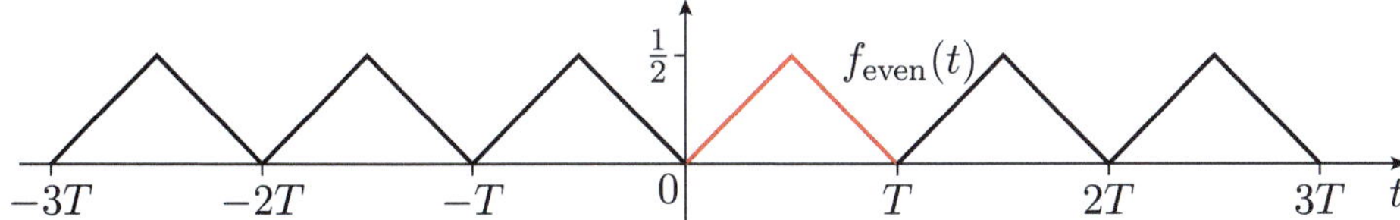

The function $f_{\text{even}}(t)$ is even and has period $\tau = T$, so its Fourier series takes the form

$$F_{\text{even}}(t) = A_0 + \sum_{n=1}^{\infty} A_n \cos\left(\frac{2n\pi t}{T}\right),$$

where the Fourier coefficients A_0 and A_n are given by

$$A_0 = \frac{1}{T} \int_{-T/2}^{T/2} f_{\text{even}}(t)\, dt,$$

$$A_n = \frac{2}{T} \int_{-T/2}^{T/2} f_{\text{even}}(t) \cos\left(\frac{2n\pi t}{T}\right) dt \quad (n = 1, 2, 3, \ldots).$$

But both of the integrands are even, and on the interval $0 \leq t \leq T/2$ we have $f_{\text{even}}(t) = f(t) = t/T$, so the integrals simplify to

$$A_0 = \frac{2}{T} \int_0^{T/2} \frac{t}{T}\, dt,$$

$$A_n = \frac{4}{T} \int_0^{T/2} \frac{t}{T} \cos\left(\frac{2n\pi t}{T}\right) dt \quad (n = 1, 2, 3, \ldots).$$

Evaluating the first integral gives

$$A_0 = \frac{2}{T^2} \int_0^{T/2} t\, dt = \frac{2}{T^2} \left[\tfrac{1}{2}t^2\right]_0^{T/2} = \tfrac{1}{4}.$$

Using equation (18) with $a = 2n\pi/T$, we get

$$A_n = \frac{4}{T^2}\left(\frac{T}{2n\pi}\right)^2\left[\cos\left(\frac{2n\pi t}{T}\right) + \frac{2n\pi t}{T}\sin\left(\frac{2n\pi t}{T}\right)\right]_0^{T/2}$$

$$= \frac{(-1)^n - 1}{n^2\pi^2}.$$

So

$$F_{\text{even}}(t) = \frac{1}{4} + \frac{1}{\pi^2}\sum_{n=1}^{\infty}\frac{(-1)^n - 1}{n^2}\cos\left(\frac{2n\pi t}{T}\right).$$

Since $f(t)$ and $f_{\text{even}}(t)$ coincide on the interval $0 \le t \le T$, this is the required cosine Fourier series $F_{\text{even}}(t)$ for $f(t)$.

The first few terms in this Fourier series are

$$F_{\text{even}}(t) = \frac{1}{4} - \frac{2}{\pi^2}\left(\cos\left(\frac{2\pi t}{T}\right) + \frac{1}{3^2}\cos\left(\frac{6\pi t}{T}\right) + \frac{1}{5^2}\cos\left(\frac{10\pi t}{T}\right) + \cdots\right).$$

Solution to Exercise 26

(a) The even periodic extension is given by

$$f_{\text{even}}(t) = \begin{cases} 1 & \text{for } 0 \le t \le \frac{\pi}{2}, \\ -1 & \text{for } \frac{\pi}{2} < t \le \pi, \\ -1 & \text{for } -\pi < t < -\frac{\pi}{2}, \\ 1 & \text{for } -\frac{\pi}{2} \le t < 0, \end{cases}$$

$$f_{\text{even}}(t + 2\pi) = f_{\text{even}}(t),$$

and is drawn below.

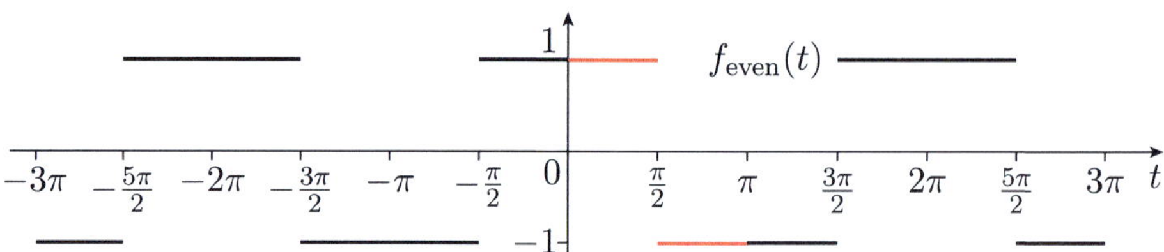

The fundamental period of this even extension is $\tau = 2\pi$, and its formula can be simplified to

$$f_{\text{even}}(t) = \begin{cases} 1 & \text{for } -\frac{1}{2}\pi \le t < \frac{\pi}{2}, \\ -1 & \text{for } \frac{\pi}{2} \le t < \frac{3\pi}{2}, \end{cases}$$

$$f_{\text{even}}(t + 2\pi) = f_{\text{even}}(t).$$

(b) Because the function $f_{\text{even}}(t)$ is even and has period $\tau = 2\pi$, its Fourier series takes the form

$$F_{\text{even}}(t) = A_0 + \sum_{n=1}^{\infty} A_n \cos nt.$$

The coefficient A_0 is given by

$$A_0 = \frac{1}{2\pi} \int_{-\pi}^{\pi} f_{\text{even}}(t)\, dt$$

$$= \frac{1}{2\pi} \left(-\int_{-\pi}^{-\pi/2} 1\, dt + \int_{-\pi/2}^{\pi/2} 1\, dt - \int_{\pi/2}^{\pi} 1\, dt \right) = 0.$$

This result is not unexpected, as the graph shows clearly that the average value of the function is zero.

As the integrand is even, the coefficients A_n are given by

$$A_n = \frac{4}{2\pi} \int_{0}^{\pi} f_{\text{even}}(t) \cos nt\, dt$$

$$= \frac{2}{\pi} \left(\int_{0}^{\pi/2} \cos nt\, dt - \int_{\pi/2}^{\pi} \cos nt\, dt \right)$$

$$= \frac{2}{n\pi} \left(\big[\sin nt\big]_0^{\pi/2} - \big[\sin nt\big]_{\pi/2}^{\pi} \right)$$

$$= \frac{4}{n\pi} \sin \frac{n\pi}{2}.$$

Hence

$$F_{\text{even}}(t) = \frac{4}{\pi} \sum_{n=1}^{\infty} \frac{1}{n} \sin \frac{n\pi}{2} \cos nt.$$

(c) The odd periodic extension is defined by

$$f_{\text{odd}}(t) = \begin{cases} 1 & \text{for } 0 < t \le \frac{\pi}{2}, \\ -1 & \text{for } \frac{\pi}{2} < t < \pi, \\ 1 & \text{for } -\pi < t < -\frac{\pi}{2}, \\ -1 & \text{for } -\frac{\pi}{2} \le t < 0, \\ 0 & \text{for } t = 0 \text{ or } t = \pi, \end{cases}$$

$$f_{\text{odd}}(t + 2\pi) = f_{\text{odd}}(t),$$

and is drawn below.

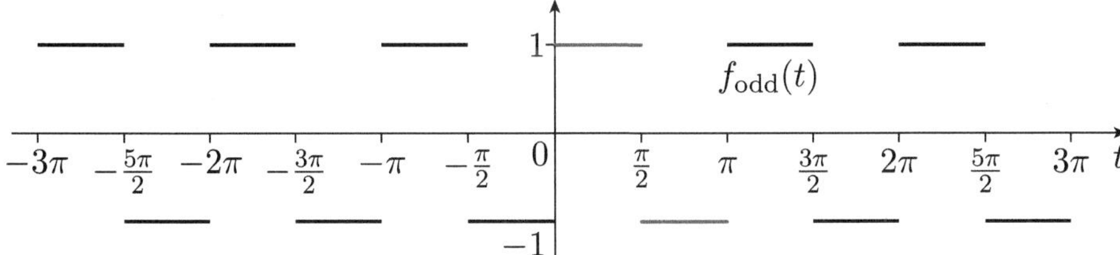

This has fundamental period $\tau = \pi$.

Partial differential equations

Introduction

A **partial differential equation** is an equation relating a dependent variable and two or more independent variables through the *partial* derivatives of the dependent variable. Differential equations have played a very important role in the module so far. But until now, all the differential equations that you have met have involved just one independent variable, and have been equations containing one or more dependent variables and their *ordinary* derivatives with respect to that independent variable. Such equations are often called *ordinary* differential equations when it is necessary to distinguish them from partial differential equations. For many systems that we want to be able to model, ordinary differential equations are inadequate because the states of the system can be specified only in terms of two – or even more – independent variables. When we are trying to model the way in which such a system changes, we are inevitably led to consider partial differential equations.

Partial derivatives were introduced in Unit 7.

Ordinary differential equations are the subject of Units 1, 6 and 12.

This unit is an introduction to partial differential equations and their solution. The method of solution introduced here is called *separation of variables*. This idea is similar to, but distinct from, the method for solving first-order ordinary differential equations that is also called separation of variables, described in Unit 1. Section 1 introduces partial differential equations and concepts associated with them, then outlines the separation of variables method that is the core of this unit.

Partial differential equations have many applications, but here we describe just two of them. In Section 2 we look at a model of the transverse vibrations of a taut string, such as a guitar string. First the model is derived in terms of a partial differential equation called the *wave equation*, then the method of separation of variables is used to find particular solutions of the wave equation, such as the vibrations of a plucked string.

The vibrations of a guitar string are also considered in Unit 11.

Section 3 looks at a different application, namely the modelling of the flow of heat in a rod. This section introduces a physical law governing heat flow called *Newton's law of cooling*, then uses this law to derive a partial differential equation modelling the flow of heat that is called the *heat equation*. The section concludes by using the separation of variables method to find particular solutions of the heat equation. A physical phenomenon known as diffusion also satisfies the same partial differential equation, so the heat equation is sometimes known as the *diffusion equation*.

1 Solving partial differential equations

This section introduces partial differential equations and describes the method of separation of variables that is used to solve them.

1.1 Introducing partial differential equations

Both the wave equation and the heat equation are examples of second-order partial differential equations, where the notion of order for partial differential equations is defined in the same way as for ordinary differential equations.

> The **order** of a partial differential equation is the order of the highest derivative that occurs.

The only issue worth highlighting here is that the order of mixed derivatives such as $\partial^2 u/\partial x\, \partial t$ is 2, because we count the total number of times that the function u is differentiated.

In this unit we consider only second-order partial differential equations, although the methods described are applicable to partial differential equations of any order. Furthermore, all the equations that we deal with are *linear*, where again linear is defined as it is for ordinary differential equations.

> A **linear** partial differential equation is one that contains no products or non-linear functions of terms involving the dependent variable and its partial derivatives.

The following exercise asks you to apply these definitions to classify some partial differential equations.

Exercise 1

We use the convention that a subscript is used to denote a partial derivative with respect to that variable. So the equation in part (d) could be written as
$$\frac{\partial^3 u}{\partial x^3} + \frac{\partial u}{\partial x} = 0.$$

For each of the following differential equations, state whether it is linear or non-linear, and write down its order.

(a) $\dfrac{\partial u}{\partial t} + u\,\dfrac{\partial u}{\partial x} = 0$ (b) $\dfrac{\partial^2 u}{\partial t^2} = x\,\dfrac{\partial u}{\partial x} + u$

(c) $\dfrac{\partial^2 u}{\partial t\, \partial x} = x^2 t$ (d) $u_{xxx} + u_x = 0$

As we will see later, the wave equation is a good model of a plucked guitar string. We will use this context to introduce other aspects of partial differential equations. A partial differential equation model is required because the state of the string – by which is meant its shape at any given time after it has been plucked – requires a function $u(x,t)$ of two independent variables, x and t, where x is the distance along the straight line joining the two points at which the string is anchored (which we can consider as an axis with origin at one end of the string), and t is the time since the string was plucked.

The straight line joining the two points at which the string is anchored is the equilibrium position of the string. The dependent variable $u = u(x, t)$ is the transverse displacement of the string from the point on the axis determined by x, at time t. For fixed $t = t_1$ and varying x, $u(x, t_1)$ specifies the shape of the string at time t_1, as shown in Figure 1(a). On the other hand, we can think about a fixed $x = x_1$ and varying t, and then $u(x_1, t)$ tells us how the transverse displacement of the string from the fixed point $x = x_1$ on the axis varies with time (see Figure 1(b)).

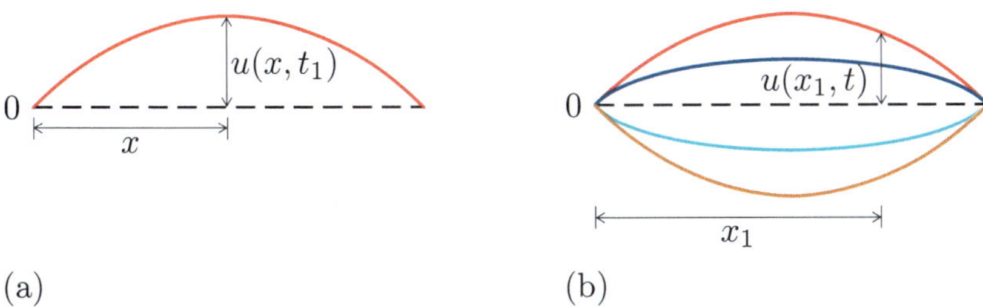

Figure 1   Motion of a plucked string: (a) string position with t fixed as t_1, and x varying; (b) string position with x fixed as x_1, and t varying – four snapshots in time are shown in different colours

The model of the motion of the string that we will develop will be a differential equation for the variable u. Since u depends on both x and t, an equation that models the motion of the string will involve partial derivatives of u with respect to both x and t. One such equation is

$$\frac{\partial^2 u}{\partial x^2} = \frac{1}{c^2} \frac{\partial^2 u}{\partial t^2}, \tag{1}$$

where c is a constant whose value depends on various physical characteristics of the string. This partial differential equation is called the **wave equation**. It is a very important equation of mathematical physics, in part because it occurs in many situations that involve vibrations of extended flexible objects like strings and springs, or indeed other wave motions such as sound waves and light waves.

The name *wave* equation is used because it models wave-like motions such as that of a plucked guitar string.

Checking whether or not a given function is a solution of a given partial differential equation is simply a matter of substituting the function into the equation, and seeing whether it is satisfied. The only difference from the case of an ordinary differential equation is that you have to calculate all the relevant partial derivatives.

Example 1

Check that

$$u(x, t) = \sin(kx) \cos(kct)$$

is a solution of the wave equation (1) for any constant k.

Solution

We have

$$\frac{\partial u}{\partial x} = k\cos(kx)\cos(kct), \quad \frac{\partial^2 u}{\partial x^2} = -k^2 \sin(kx)\cos(kct),$$

$$\frac{\partial u}{\partial t} = -kc\sin(kx)\sin(kct), \quad \frac{\partial^2 u}{\partial t^2} = -k^2 c^2 \sin(kx)\cos(kct).$$

So when $u(x,t) = \sin(kx)\cos(kct)$,

$$\frac{\partial^2 u}{\partial x^2} = \frac{1}{c^2}\frac{\partial^2 u}{\partial t^2},$$

and this function is indeed a solution of the wave equation.

Now try this yourself by attempting the following exercise.

Exercise 2

Check that $u(x,t) = \sin(x)\,e^{-\alpha t}$ is a solution of the partial differential equation

$$\frac{\partial u}{\partial t} = \alpha \frac{\partial^2 u}{\partial x^2}. \tag{2}$$

This partial differential equation is known as the *heat equation*.

1.2 Initial conditions and boundary conditions

Initial conditions and boundary conditions for partial differential equations play the same role as they do for ordinary differential equations: they can be applied to a general solution to get a particular solution.

The initial conditions and/or boundary conditions appropriate to obtaining a particular solution of the wave equation or the heat equation depend on the context. For example, for the wave equation model of the vibrations of a guitar string, we need to use *two* boundary conditions and *two* initial conditions in order to have a unique solution. This is because the wave equation involves the second partial derivative with respect to x (hence the need for two boundary conditions) and also involves the second partial derivative with respect to t (hence the need for two initial conditions). In contrast, the heat equation (2) involves only the first partial derivative with respect to t, hence it requires only one initial condition in order to have a unique solution.

For the wave equation, if L is the equilibrium length of the string, then the boundary conditions are

$$u(0,t) = 0 \quad \text{and} \quad u(L,t) = 0, \quad t \geq 0; \tag{3}$$

these conditions correspond to the string being fixed at its ends. The initial conditions model the action of plucking the string, which sets it in motion. Plucking consists of holding the string in a certain shape, at rest, and then releasing it. If the initial shape of the string is given by a function $f(x)$, then the initial conditions may be specified in the form

$$u(x,0) = f(x), \quad 0 < x < L,$$
$$u_t(x,0) = 0, \quad 0 \le x \le L.$$

The first initial condition models the initial shape of the string, while the second corresponds to it being at rest initially. The initial condition for a particular initial shape is given in the following example.

We use the range $0 < x < L$ for the initial condition $u(x,0) = f(x)$ because the values of $u(x,0)$ when $x = 0$ and $x = L$ are specified by boundary conditions (3).

Example 2

A taut string of equilibrium length L is plucked at its midpoint, which is given an initial displacement $\frac{1}{2}$, as shown in Figure 2. It is then released from rest.

Write down the initial conditions for the wave equation for the transverse vibrations of this string.

Solution

The displacement shown in Figure 2 has two linear sections, with slopes $\pm 1/L$. Hence the initial displacement is given by

$$u(x,0) = \begin{cases} x/L & \text{for } 0 < x \le \frac{1}{2}L, \\ (L-x)/L & \text{for } \frac{1}{2}L < x < L. \end{cases}$$

As the string is released from rest, the transverse component of the initial velocity is given by

$$u_t(x,0) = 0, \quad 0 \le x \le L.$$

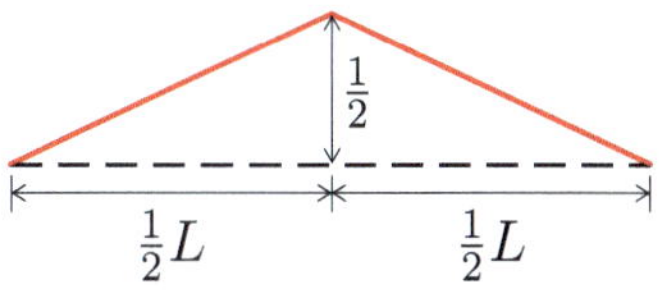

Figure 2 Initial position of a taut string plucked at its midpoint

Exercise 3

Suppose that the initial conditions for the transverse vibrations of a taut string are

$$u(x,0) = \begin{cases} -\dfrac{4d}{L}x & \text{for } 0 < x \le \frac{1}{4}L, \\ -\dfrac{4d}{3L}(L-x) & \text{for } \frac{1}{4}L < x < L, \end{cases}$$
$$u_t(x,0) = 0, \quad 0 \le x \le L.$$

Describe how the string has been set in motion.

Exercise 4

A taut string of equilibrium length L is plucked one-third of the way along its length, which is given an initial displacement d, as shown in Figure 3.

What is the corresponding initial condition, for this displacement, for the wave equation for the transverse vibrations of this string?

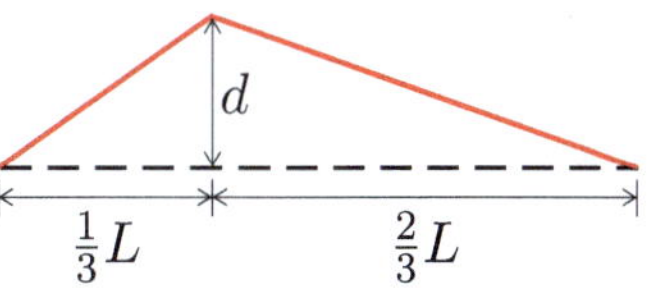

Figure 3 Initial position of a taut string plucked at a point a third of the way along its length

Exercise 5

A taut string is initially in its equilibrium position. At time $t = 0$, it is struck in such a way as to impart, instantaneously, a transverse velocity v (in the positive direction) to the middle third of the string. This is a simple model of a string that is hammered, such as a piano wire.

Modify the initial conditions for the wave equation for transverse vibrations of the string to model this situation.

Example 1 showed that any function of the form $u(x, t) = \sin(kx)\cos(kct)$ is a solution of the wave equation, for any constant k. In particular, therefore,

$$u(x, t) = \sin\left(\frac{\pi x}{L}\right)\cos\left(\frac{\pi ct}{L}\right) \tag{4}$$

is a solution of the wave equation for transverse vibrations of a taut string, where L is the equilibrium length of the string. The following exercise asks you to show that this solution also satisfies fixed endpoint boundary conditions and the initial condition that the string starts from rest.

Exercise 6

Show that the solution given by equation (4) satisfies the boundary conditions

$$u(0, t) = u(L, t) = 0, \quad t \geq 0,$$

and the initial condition

$$u_t(x, 0) = 0, \quad 0 \leq x \leq L.$$

Equation (4) is not the only solution of the wave equation that satisfies the boundary conditions. You may like to verify that each member of the following family of functions $u_n(x, t)$ $(n = 1, 2, \ldots)$ also satisfies the wave equation, the fixed endpoint boundary conditions and the starting from rest initial condition:

$$u_n(x, t) = \sin\left(\frac{\pi n x}{L}\right)\cos\left(\frac{\pi n ct}{L}\right). \tag{5}$$

The first three members of this family of solutions are shown in Figure 4 at time $t = 0$.

(Note that $u_n(x, t)$ denotes a family of functions indexed by the discrete variable n; it is not the derivative with respect to n, as the notation would imply if n were a continuous variable.)

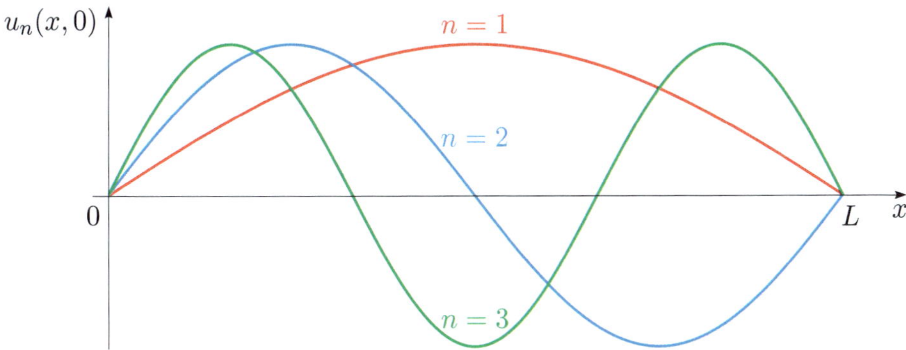

Figure 4   The first three transverse vibrations of a taut string

You may recognise the waves shown in Figure 4 from earlier physics or music theory as having the shape of the fundamental $(n = 1)$ and harmonics $(n = 2, 3, \ldots)$ of a vibrating string. You can observe that the larger the value of n, the higher the frequency (and the higher the pitch of the note).

In music we rarely hear pure tones such as the ones shown in Figure 4 as musical instruments each produce a characteristic mixture of pure tones that gives the characteristic sound (known as the timbre) of an instrument. In terms of a mathematical model of a taut string, the mixing of tones corresponds to taking linear combinations of the family of solutions (5). This is a key feature of the type of partial differential equations that we study, namely linear differential equations that are **homogeneous**, in that each additive term involves the dependent variable or its derivatives – there are no constant terms or terms involving solely the independent variables. The key result is as follows.

Principle of superposition

If u and v are solutions of a linear homogeneous partial differential equation, then $Au + Bv$ is also a solution of the same equation for any constants A and B. Furthermore, if u and v also satisfy homogeneous boundary conditions such as $u(0, t) = 0$ or $u_x(L, t) = 0$, then the linear combination $Au + Bv$ will also satisfy them.

The principle of superposition enables the construction of a general solution of a partial differential equation from particular solutions such as the family of solutions (5).

It may seem that this principle of superposition is of limited applicability since it applies only to homogeneous boundary conditions. The following exercises show that this is not the case, as inhomogeneous boundary conditions can be reduced to homogeneous boundary conditions.

Exercise 7

Consider the heat equation

$$\frac{\partial \Theta}{\partial t} = \alpha \, \frac{\partial^2 \Theta}{\partial x^2},$$

subject to boundary conditions

$$\Theta(0,t) = \Theta(L,t) = \Theta_0, \quad t \geq 0,$$

where $\Theta_0 \neq 0$.

(a) Show that the function $\Theta(x,t) = \Theta_0$ satisfies the differential equation and boundary conditions.

(b) Define $u(x,t) = \Theta(x,t) - \Theta_0$. Show that $u(x,t)$ satisfies the heat equation with homogeneous boundary conditions.

Exercise 8

Consider the heat equation

$$\frac{\partial \Theta}{\partial t} = \alpha \, \frac{\partial^2 \Theta}{\partial x^2},$$

subject to boundary conditions

$$\Theta(0,t) = \Theta_0, \quad \Theta(L,t) = \Theta_L, \quad t \geq 0,$$

where Θ_0 and Θ_L are non-zero constants.

(a) Show that the function

$$\Theta(x,t) = \frac{L-x}{L}\Theta_0 + \frac{x}{L}\Theta_L$$

satisfies the differential equation and the boundary conditions.

(b) Define the function

$$u(x,t) = \Theta(x,t) - \frac{L-x}{L}\Theta_0 - \frac{x}{L}\Theta_L.$$

Show that $u(x,t)$ satisfies the heat equation with homogeneous boundary conditions.

Results similar to those in Exercises 7 and 8 allow us to concentrate on the case of homogeneous boundary conditions, which we will mainly consider from now on.

The case of homogeneous boundary conditions is simpler because we can then apply the principle of superposition. This will be the strategy for solving partial differential equations: first find a family of particular solutions, then use the principle of superposition to find the general solution. The next subsection describes this strategy.

1.3 Separation of variables

This subsection is the core of this unit. It describes a method known as separation of variables, which is one of the few techniques available for solving linear partial differential equations; it is the only technique that we will study in this unit. As mentioned in the previous subsection, the strategy is to look for particular solutions that can be combined to give the general solution. We look for solutions of the form

$$u(x,t) = X(x)\,T(t).$$

We now have to find the relevant partial derivatives of the function $u(x,t)$ in terms of the functions $X(x)$ and $T(t)$.

Exercise 9

If the function u is defined as a product

$$u(x,t) = X(x)\,T(t),$$

find formulas for the partial derivatives $\partial u/\partial t$, $\partial^2 u/\partial t^2$, $\partial u/\partial x$ and $\partial^2 u/\partial x^2$ in terms of the functions X and T and their ordinary derivatives.

Using the derivatives obtained in Exercise 9, we can separate variables in a partial differential equation in a similar (but distinct) way to the separation of variables method for solving first-order differential equations that you studied in Unit 1. As an example, consider the equation

$$\frac{\partial^2 u}{\partial x^2} = \frac{\partial u}{\partial t}, \tag{6}$$

and look for solutions of the form $u(x,t) = X(x)\,T(t)$. From Exercise 9 we have

$$\frac{\partial^2 u}{\partial x^2} = X''(x)\,T(t) \quad \text{and} \quad \frac{\partial u}{\partial t} = X(x)\,T'(t).$$

Substituting into the partial differential equation gives

$$X''(x)\,T(t) = X(x)\,T'(t).$$

Dividing by $X(x)\,T(t)$ gives

$$\frac{X''(x)}{X(x)} = \frac{T'(t)}{T(t)}.$$

This has achieved our aim of separating the variables, as the left-hand side involves only x and the right-hand side involves only t. This equation must hold for all x and t, and the only way that this can happen is if both sides are equal to the same constant. Let μ be this constant, which is called the **separation constant**. So we have

$$\frac{X''(x)}{X(x)} = \mu \quad \text{and} \quad \frac{T'(t)}{T(t)} = \mu.$$

Rearranging these equations gives two ordinary differential equations:

$$X''(x) - \mu\, X(x) = 0 \quad \text{and} \quad T'(t) - \mu\, T(t) = 0. \tag{7}$$

Each of these equations can be solved separately by using the methods of Unit 1, but before we progress to doing this, try the following exercise to separate the variables for another partial differential equation.

Exercise 10

Consider the differential equation

$$\frac{\partial^2 u}{\partial x^2} + 2u = \frac{\partial u}{\partial t}.$$

Use the substitution $u(x,t) = X(x)\,T(t)$ to separate the variables x and t, and hence find the two differential equations satisfied by $X(x)$ and $T(t)$.

Ordinary differential equations similar to equations (7) will occur frequently for the partial differential equations that we consider in this unit. These equations can be solved by using the methods of Unit 1, and for reference we recall the solutions here.

The general solution of the equation $T'(t) - \mu\, T(t) = 0$ is

$$T(t) = C \exp(\mu t), \tag{8}$$

where C is a constant.

The form of the solution of the equation involving x depends on the value of the separation constant μ and splits into three cases.

The general solution of the equation $X''(x) - \mu\, X(x) = 0$ is

$$X(x) = \begin{cases} Ae^{cx} + Be^{-cx} & \text{for } \mu > 0, \\ Ax + B & \text{for } \mu = 0, \\ A\cos kx + B\sin kx & \text{for } \mu < 0, \end{cases} \tag{9}$$

where A and B are constants, $c = \sqrt{\mu}$ and $k = \sqrt{-\mu}$.

Note that the constants c and k that appear in the box above are real numbers (in the cases where they apply) and are positive (since the square root function gives the positive root).

Now we turn our attention to the boundary conditions for the partial differential equation, which can be used to derive boundary conditions for the separated ordinary differential equations. As an example to show the general method, consider the boundary condition $u(0,t) = 0$ for $t \geq 0$. Substituting $u(x,t) = X(x)\,T(t)$ into the boundary condition gives

$$X(0)\,T(t) = 0.$$

This equation implies that either $X(0) = 0$ or $T(t) = 0$ for all t. The latter option gives $u(x,t) = X(x) \times 0 = 0$, which is known as the *trivial* solution (it is always a solution of a linear homogeneous differential equation). So for non-trivial solutions of the partial differential equation we must have $X(0) = 0$. Hence the boundary condition for $u(x,t)$ imposes a boundary condition on $X(x)$.

Similar results hold for other boundary conditions, as the next exercise shows.

Exercise 11

Consider the boundary condition $u_x(1, t) = 0$ for $t \geq 0$. If $u(x, t) = X(x)\,T(t)$, then what boundary condition must $X(x)$ satisfy in order to find non-trivial solutions of a partial differential equation?

Returning now to the differential equation (6), namely $u_{xx} = u_t$, we consider the solutions subject to the boundary conditions $u(0, t) = u(1, t) = 0$ for $t \geq 0$. In terms of X and T, the boundary conditions become $X(0)\,T(t) = 0$ and $X(1)\,T(t) = 0$, hence for non-trivial solutions we must have $X(0) = 0$ and $X(1) = 0$.

The aim is now to find non-trivial solutions of the differential equation $X''(x) - \mu\,X(x) = 0$ that satisfy the boundary conditions. In equation (9) there are three different forms of the solution depending on the sign of μ, and we consider each in turn.

- $\mu > 0$. Let $c = \sqrt{\mu}$. As stated in equation (9), the general solution in this case is

$$X(x) = Ae^{cx} + Be^{-cx},$$

  where A and B are constants.

  The boundary condition $X(0) = 0$ gives the equation $Ae^0 + Be^0 = 0$, that is, $A + B = 0$ or $B = -A$.

  The boundary condition $X(1) = 0$ gives the equation

$$Ae^c + Be^{-c} = 0.$$

  As $B = -A$, this simplifies to

$$A(e^c - e^{-c}) = 0.$$

  Multiplying both sides by e^c gives

$$A(e^{2c} - 1) = 0.$$

  Now we argue that the term in brackets is never zero. Since $c > 0$, we have $2c > 0$, and taking exponentials gives $\exp(2c) > \exp(0) = 1$ (this last step is a consequence of the fact that e^x is an increasing function). Thus the term $(e^{2c} - 1)$ is never zero, and we conclude that $A = 0$ and $B = -A = 0$.

  So the only solution is the trivial solution $X(x) = 0$.

- $\mu = 0$. As stated in equation (9), the general solution in this case is

$$X(x) = Ax + B,$$

where A and B are constants.

The boundary condition $X(0) = 0$ gives $B = 0$. The boundary condition $X(1) = 0$ then gives $A = 0$.

So the only solution is the trivial solution $X(x) = 0$.

Note that as μ is negative, $\sqrt{-\mu}$ is a positive real number.

- $\mu < 0$. Let $k = \sqrt{-\mu}$. As stated in equation (9), the general solution can be written as

$$X(x) = A \cos kx + B \sin kx,$$

where A and B are constants.

The boundary condition $X(0) = 0$ yields $A \cos 0 + B \sin 0 = 0$, so $A = 0$. The boundary condition $X(1) = 0$ then gives $B \sin k = 0$. So either $B = 0$ or $\sin k = 0$. The option $B = 0$ leads to $X(x) = 0$ again, so for non-trivial solutions we must have

$$\sin k = 0.$$

The zeros of the sine function occur at integer multiples of π, so $k = n\pi$ for some integer n.

So the function

$$X(x) = B \sin(n\pi x) \quad \text{for } n = 1, 2, \ldots$$

Note that n must be positive since k is positive.

is a solution of the differential equation that satisfies the boundary condition for any positive integer n. The separation constant for this solution is $\mu = -k^2 = -n^2\pi^2$.

Now we have found non-trivial solutions for $X(x)$, we proceed to find the corresponding solutions for $T(t)$.

Recall that the differential equation for $T(t)$ is $T'(t) - \mu\, T(t) = 0$, which becomes

$$T'(t) + n^2\pi^2\, T(t) = 0$$

as $\mu = -n^2\pi^2$. From equation (8), the general solution of this equation is

$$T(t) = C \exp(-n^2\pi^2 t),$$

where C is a constant.

Multiplying the solutions for $X(x)$ and $T(t)$ gives a function $u(x,t)$ that satisfies the partial differential equation and boundary conditions:

$$u(x,t) = a \sin(n\pi x) \exp(-n^2\pi^2 t),$$

where we have combined the two constants into one: that is, $a = BC$.

There is one solution for each positive integer n, so we have a family of solutions

$$u_n(x,t) = a_n \sin(n\pi x)\exp(-n^2\pi^2 t) \quad \text{for } n = 1, 2, \ldots,$$

where we have added a subscript n to the constant a to emphasise that these constants can have a different value for each value of n. These solutions of the partial differential equation are known as **normal mode solutions**.

As the partial differential equation and boundary conditions are homogeneous and linear, any superposition of these solutions is also a solution. So we can write the general solution as

$$u(x,t) = \sum_{n=1}^{\infty} a_n \sin(n\pi x)\exp(-n^2\pi^2 t).$$

To determine the constants, we need an initial condition for the partial differential equation, such as

$$u(x,0) = \sin(2\pi x).$$

Substituting the general solution into this equation gives

$$\sum_{n=1}^{\infty} a_n \sin(n\pi x) = \sin(2\pi x). \tag{10}$$

You should recognise this as a Fourier sine series that would result from computing the Fourier series of the odd periodic extension of $\sin(2\pi x)$ defined on the interval $[0,1]$. We can proceed, as in Unit 13, to determine the coefficients by integration:

$$a_n = \frac{2}{2\pi} \int_{-\pi}^{\pi} \sin(2\pi x)\sin(n\pi x)\,dx.$$

This integral has already been evaluated in Unit 13 – the integral over a complete period of a product of sines is zero unless the arguments are the same. So we have $a_2 = 1$, and $a_n = 0$ if $n \neq 2$. In fact, this result could have been written down 'by inspection' of equation (10), as the function on the right-hand side is one of the terms of the series on the left-hand side; equation (10) could be written as

$$a_1 \sin(\pi x) + a_2 \sin(2\pi x) + a_3 \sin(3\pi x) + \cdots = \sin(2\pi x).$$

So the particular solution that satisfies the given initial condition is

$$u(x,t) = \sin(2\pi x)\exp(-4\pi^2 t).$$

In finding this solution we have gone through several steps, and it is worthwhile summarising them now in the form of a procedure. The terms in the margin will be used in later examples to show how this procedure is being applied.

Procedure 1   Separation of variables

To find the solution of a homogeneous linear partial differential equation with dependent variable u and independent variables x and t, subject to boundary and initial conditions, carry out the following steps.

1. Separate the variables by substituting the trial solution

$$u(x,t) = X(x)\,T(t)$$

into the partial differential equation and rearranging so that each side of the equation involves only one of the variables. Both sides must then be equal to a separation constant μ.

Rearranging then gives two separate ordinary differential equations for X and T. The boundary conditions for u will give boundary conditions for X.

2. Find the general solution of the ordinary differential equations for X and T found in Step 1. Use the boundary conditions for X to find the normal mode solutions $u_n(x,t)$.

3. Write down the general solution as a linear combination of the normal mode solutions:

$$u(x,t) = \sum_{n=0}^{\infty} a_n\, u_n(x,t).$$

The initial conditions (and results about Fourier series) can be used to determine the constants a_n appearing in this solution.

◀ Separate variables ▶

◀ Solve ODEs ▶

◀ Initial conditions ▶

The next exercise asks you to follow the first two steps of this procedure to find a family of normal mode solutions for a different partial differential equation.

Exercise 12

Use Procedure 1 to find an infinite family of normal mode solutions for the partial differential equation

$$\frac{\partial^2 u}{\partial x^2} + \frac{\partial^2 u}{\partial t^2} = 0,$$

This equation is known as *Laplace's equation.*

subject to boundary conditions

$$u(0,t) = u(1,t) = 0, \quad t \geq 0.$$

Procedure 1 is the key technique introduced in this unit. The remaining sections apply this technique to analyse models in two different contexts: modelling the vibrations of a string and modelling the flow of heat. The next section concentrates on the first of these.

2 The wave equation

This section is devoted to deriving and solving the wave equation. In Subsection 2.1 we will see how the wave equation arises as a model of the transverse vibrations of a taut string, and also how damping can be incorporated into the model. Subsection 2.2 looks at modelling the vibrations of a plucked string. The model for this is the wave equation subject to fixed endpoint boundary conditions and initial conditions specifying a given initial string profile and zero initial velocity. Subsection 2.3 looks at what happens when damping is incorporated into the model.

2.1   Deriving the wave equation

In this subsection you will see how a continuous model of a guitar string – or indeed of any taut string – leads to a model of the transverse vibrations of the string as a second-order partial differential equation. This derivation is given for completeness – you will not be asked to reproduce or modify this derivation in the assessment of this module.

You saw a *discrete* model of a guitar string in Section 3 of Unit 11.

The assumptions needed to develop this model are as follows.

(a) The string is uniform, with total mass M and equilibrium length L.

(b) Each point of the string is subject only to a small, smoothly varying transverse displacement u, as shown in Figure 5.

(c) All external forces, such as weight, friction and air resistance, are negligible in comparison with the tension in the string.

Using assumption (c) that the weight of the string is negligible in comparison to the tension in the string implies that when it is in equilibrium, the string lies in a straight line. We will take this equilibrium line as the x-axis, with origin at the left-hand end; see Figure 5.

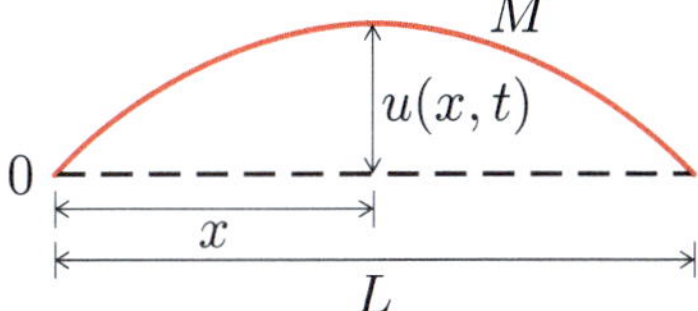

Figure 5   A taut string

Consider a small segment of length δx a distance x along the string (so that in equilibrium it would occupy the interval $[x - \delta x/2, x + \delta x/2]$), as shown in Figure 6. Also shown in the figure are the two neighbouring segments.

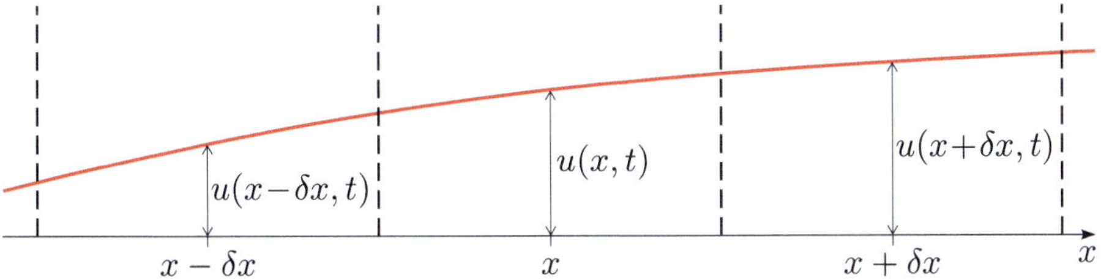

Figure 6   Splitting the string into small segments: imagine the string being split at the dashed lines and replaced by particles located at the midpoints

By assumption (a) the string is uniform, so the mass of each segment is $M\,\delta x/L$. So each small segment can be modelled as a particle of mass

$M\,\delta x/L$ located at the point $(x, u(x,t))$, as shown in Figure 7. Also shown in Figure 7 is the angle $\theta(x,t)$, which is the angle that the tension force due to the string makes to the left of the particle at position x at time t.

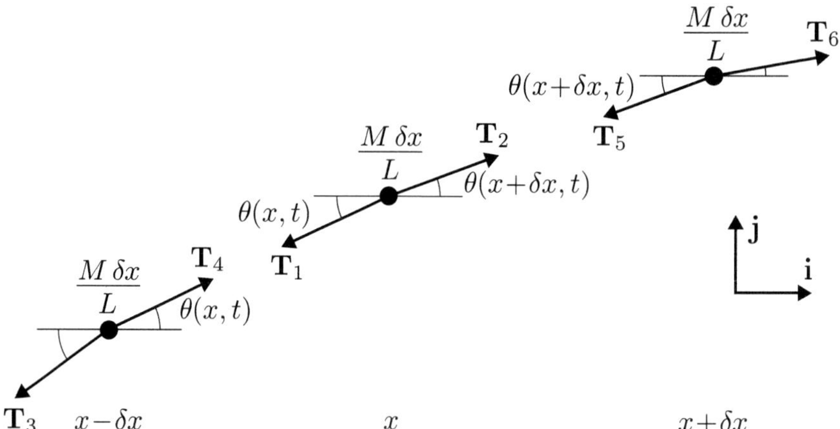

Figure 7   Force diagram for three neighbouring elements of string (vertical scale exaggerated for clarity)

Also shown in Figure 7 are the forces acting on the three segments. By assumption (c), all forces are negligible compared to the tension in the string, so we have modelled only the tension forces. As the tension forces between neighbouring particles (e.g. $\mathbf{T}_4$ and $\mathbf{T}_1$) are due to a portion of string at a fixed angle, they must be of equal magnitude and opposite direction, as shown. The total force on the central segment at position x is $\mathbf{T}_1 + \mathbf{T}_2$.

By assumption (b), the central segment moves vertically, so its acceleration is $u_{tt}(x,t)\,\mathbf{j}$. Applying Newton's second law gives

$$\frac{M\,\delta x}{L}\,u_{tt}(x,t)\,\mathbf{j} = \mathbf{T}_1 + \mathbf{T}_2. \tag{11}$$

We now need expressions for $\mathbf{T}_1$ and $\mathbf{T}_2$. Resolving in the $\mathbf{i}$-direction gives

$$0 = \mathbf{T}_1 \cdot \mathbf{i} + \mathbf{T}_2 \cdot \mathbf{i}.$$

From this equation we get $|\mathbf{T}_1 \cdot \mathbf{i}| = |\mathbf{T}_2 \cdot \mathbf{i}|$, that is, the magnitude of the horizontal component of the tension force is constant. As the string is horizontal in equilibrium, this is equal to the equilibrium tension, T_{eq}. This gives the diagram shown in Figure 8.

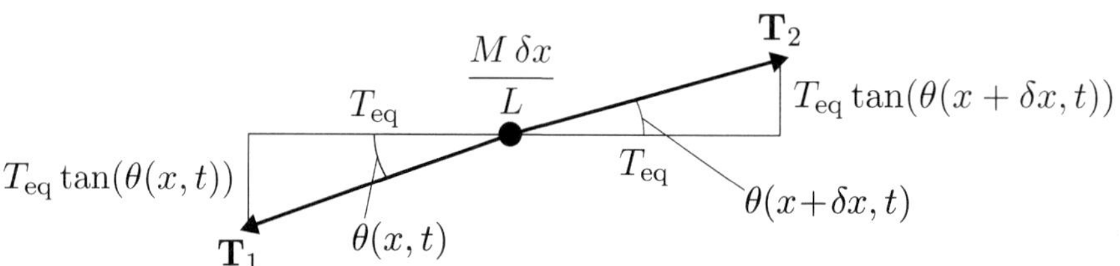

Figure 8   Resolving the forces in the central line segment (vertical scale exaggerated for clarity)

From Figure 8 we have

$$\mathbf{T}_1 = -T_{\text{eq}}\mathbf{i} - T_{\text{eq}}\tan(\theta(x,t))\,\mathbf{j}$$

and

$$\mathbf{T}_2 = T_{\text{eq}}\mathbf{i} + T_{\text{eq}}\tan(\theta(x+\delta x,t))\,\mathbf{j}.$$

What remains to be done is to relate the angle $\theta(x,t)$ to the string displacement $u(x,t)$. From the triangle shown in Figure 9, we obtain

$$\tan(\theta(x,t)) = \frac{u(x,t) - u(x - \delta x,t)}{\delta x}.$$

As δx becomes smaller, the right-hand side of this equation tends to the first derivative of the displacement with respect to x, that is, $u_x(x,t)$. So $\tan(\theta(x,t)) = u_x(x,t)$, and similarly $\tan(\theta(x+\delta x,t)) = u_x(x+\delta x,t)$, hence we have

$$\mathbf{T}_1 = -T_{\text{eq}}\mathbf{i} - T_{\text{eq}}\,u_x(x,t)\,\mathbf{j}$$

and

$$\mathbf{T}_2 = T_{\text{eq}}\mathbf{i} + T_{\text{eq}}\,u_x(x+\delta x,t)\,\mathbf{j}.$$

Substituting into equation (11) and resolving in the $\mathbf{j}$-direction gives

$$\frac{M\,\delta x}{L}\,u_{tt}(x,t) = T_{\text{eq}}\,u_x(x+\delta x,t) - T_{\text{eq}}\,u_x(x,t). \tag{12}$$

Rearranging yields

$$\frac{M}{T_{\text{eq}}L}\,u_{tt}(x,t) = \frac{u_x(x+\delta x,t) - u_x(x,t)}{\delta x}.$$

Taking the limit as $\delta x \to 0$ gives

$$\frac{M}{T_{\text{eq}}L}\,u_{tt}(x,t) = u_{xx}(x,t).$$

This is the differential equation that we were aiming to derive, but it is more usual to write

$$c^2 = T_{\text{eq}}L/M \tag{13}$$

and present the equation as follows.

Wave equation

$$\frac{\partial^2 u}{\partial x^2} = \frac{1}{c^2}\frac{\partial^2 u}{\partial t^2}. \tag{14}$$

Example 3

Use equation (14) to determine the dimensions of the constant c.

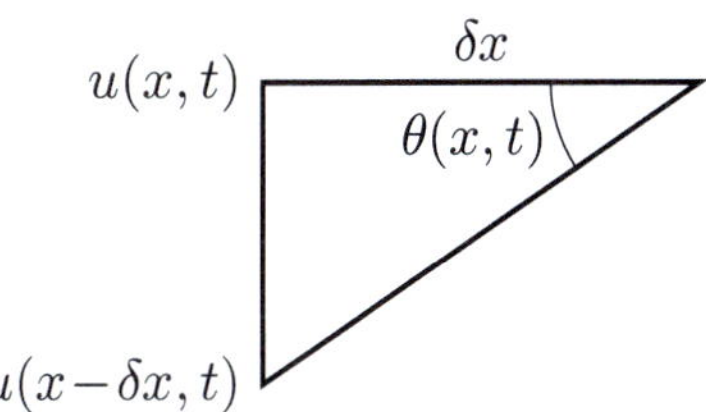

Figure 9   Determining the relationship between $\theta(x,t)$ and $u(x,t)$

Recall from Unit 7 that

$$\lim_{h\to 0}\frac{f(x+h,t)-f(x,t)}{h} = f_x(x,t).$$

Here $f = u$ and $h = -\delta x$.

Now in

$$\lim_{h\to 0}\frac{f(x+h,t)-f(x,t)}{h} = f_x(x,t)$$

we have $f = u_x$ and $h = \delta x$.

Solution

The dimensions of a derivative were dealt with in Exercise 21 of Unit 8.

$$[\partial^2 u/\partial x^2] = \mathrm{L}\,\mathrm{L}^{-2} = \mathrm{L}^{-1},$$
$$[\partial^2 u/\partial t^2] = \mathrm{L}\,\mathrm{T}^{-2}.$$

Therefore

$$[c^2] = \frac{[\partial^2 u/\partial t^2]}{[\partial^2 u/\partial x^2]} = \frac{\mathrm{L}\,\mathrm{T}^{-2}}{\mathrm{L}^{-1}} = \mathrm{L}^2\mathrm{T}^{-2},$$

so $[c] = \mathrm{L}\,\mathrm{T}^{-1}$.

So c has the dimensions of speed, and equation (14) will be dimensionally consistent provided that $\sqrt{T_{\mathrm{eq}}L/M}$ has the same dimensions.

Exercise 13

Show that $\sqrt{T_{\mathrm{eq}}L/M}$ has the dimensions of speed.

This subsection concludes by deriving a more complicated model for the transverse vibrations of a string that includes damping. We start from equation (12), which is the equation of transverse motion of the string without damping, and now consider adding a linear damping force to the motion. A first model of the damping force assumes that its magnitude is proportional to the length of the segment δx and to the transverse component of the segment's velocity $u_t(x,t)$, and that it acts in the direction opposite to the velocity. Equation (12) then becomes

$$\frac{M\,\delta x}{L}\,u_{tt}(x,t) = T_{\mathrm{eq}}\,u_x(x+\delta x,t) - T_{\mathrm{eq}}\,u_x(x,t) - \alpha\,\delta x\,u_t(x,t),$$

where α is a constant to be determined experimentally. Dividing through by δx and taking the limit as $\delta x \to 0$, we obtain

$$\frac{M}{L}\frac{\partial^2 u}{\partial t^2}(x,t) = T_{\mathrm{eq}}\frac{\partial^2 u}{\partial x^2}(x,t) - \alpha\frac{\partial u}{\partial t}(x,t).$$

This is the partial differential equation for damped transverse oscillations, but it is more usually written with different parameters by writing $c^2 = T_{\mathrm{eq}}L/M$ and $\varepsilon = \alpha L/(2M)$ to obtain the following equation.

Damped wave equation

$$\frac{\partial^2 u}{\partial x^2} = \frac{1}{c^2}\left(\frac{\partial^2 u}{\partial t^2} + 2\varepsilon\frac{\partial u}{\partial t}\right). \tag{15}$$

2.2 Solving the wave equation

In this subsection we will solve the wave equation for vibrations of a taut string.

Example 4

In this example we consider the vibrations of a guitar string of length L that is held fixed at its endpoints. The string is plucked, that is, released from rest with an initial profile given by a function $f(x)$, $0 < x < L$.

We model the guitar string as a taut string, so the transverse displacement of the string at position x and time t will satisfy the wave equation

$$\frac{\partial^2 u}{\partial x^2} = \frac{1}{c^2}\frac{\partial^2 u}{\partial t^2}. \tag{16}$$

The fixed endpoints of the string give the boundary conditions

$$u(0,t) = 0 \quad \text{and} \quad u(L,t) = 0, \quad t \geq 0. \tag{17}$$

The initial profile of the string gives the initial condition

$$u(x,0) = f(x), \quad 0 < x < L, \tag{18}$$

and the fact that it is released from rest gives the initial condition

$$u_t(x,0) = 0, \quad 0 \leq x \leq L. \tag{19}$$

Find $u(x,t)$ that satisfies this model.

Solution

Applying Procedure 1, we look for solutions of the form

$$u(x,t) = X(x)\,T(t).$$

◄ Separate variables ►

Here T is a function of time. It should not be confused with the magnitude of a force $\mathbf{T}$ or the dimension T.

Differentiating this expression gives

$$\frac{\partial u}{\partial t} = X(x)\,T'(t), \quad \frac{\partial^2 u}{\partial t^2} = X(x)\,T''(t),$$

$$\frac{\partial u}{\partial x} = X'(x)\,T(t), \quad \frac{\partial^2 u}{\partial x^2} = X''(x)\,T(t).$$

Substituting into equation (16) gives

$$X''(x)\,T(t) = \frac{1}{c^2}\,X(x)\,T''(t),$$

which on rearranging gives

$$\frac{X''(x)}{X(x)} = \frac{T''(t)}{c^2\,T(t)}.$$

Since the left-hand side is a function of x only, and the right-hand side is a function of t only, both sides must be constant and equal to a separation constant μ. This gives the two equations

$$\frac{X''(x)}{X(x)} = \mu \quad \text{and} \quad \frac{T''(t)}{c^2\,T(t)} = \mu.$$

Rearranging gives two ordinary differential equations:

$$X''(x) = \mu\,X(x), \quad T''(t) = \mu c^2\,T(t).$$

The given boundary conditions (17) become

$$X(0)\,T(t) = 0 \quad \text{and} \quad X(L)\,T(t) = 0, \quad t \geq 0.$$

One solution of these equations is $T(t) = 0$ for $t \geq 0$, which leads to the trivial solution $u(x, t) = 0$ that is always a possibility. For non-trivial solutions $T(t)$ is not always zero, so we must have

$$X(0) = 0 \quad \text{and} \quad X(L) = 0.$$

◀ Solve ODEs ▶

Now we consider solving the differential equation for X. The general solution is given by equation (9) and splits into three cases depending on the sign of μ.

- $\mu > 0$. The general solution can be written as

$$X(x) = Ae^{cx} + Be^{-cx},$$

where $c = \sqrt{\mu}$, and A and B are constants. The boundary condition $X(0) = 0$ gives $A + B = 0$, that is, $B = -A$. The boundary condition $X(L) = 0$ then gives

$$A(e^{cL} - e^{-cL}) = 0.$$

Multiplying by e^{cL} yields

$$A(e^{2cL} - 1) = 0.$$

Now we can argue as before that the term in brackets is never zero. As $2cL > 0$, we must have $\exp(2cL) > \exp(0) = 1$ as the exponential function is everywhere increasing. So $A = 0$ and the only solution in this case is the trivial solution $X(x) = 0$.

- $\mu = 0$. Now equation (9) gives the general solution as

$$X(x) = Ax + B,$$

where A and B are constants. The boundary condition $X(0) = 0$ gives $B = 0$. The boundary condition $X(L) = 0$ then gives $AL = 0$, which implies that $A = 0$ as L is non-zero (because it represents the length of the string). So the only solution in this case is the trivial solution $X(x) = 0$.

- $\mu < 0$. Let $k = \sqrt{-\mu}$. Then equation (9) gives the general solution as

$$X(x) = A \cos kx + B \sin kx,$$

where A and B are constants. The boundary condition $X(0) = 0$ gives $A = 0$. The boundary condition $X(L) = 0$ then gives

$$B \sin kL = 0.$$

So $B = 0$ or $\sin kL = 0$. The case where $B = 0$ gives the trivial solution $X(x) = 0$ for all x. So for non-trivial solutions we must have $\sin kL = 0$, which gives $kL = n\pi$ for some positive integer n, that is,

$$k = \frac{n\pi}{L}, \quad n = 1, 2, 3, \ldots.$$

So the non-trivial solution is

$$X(x) = B \sin\left(\frac{n\pi x}{L}\right), \quad n = 1, 2, 3, \ldots. \tag{20}$$

In this case the separation constant μ is given by

$$\mu = -k^2 = -\frac{n^2\pi^2}{L^2}.$$

We now turn to the function $T(t)$. Substituting for the value of μ, the differential equation for T becomes

$$T''(t) + \frac{c^2 n^2 \pi^2}{L^2} T(t) = 0.$$

The general solution of this equation is stated in equation (9) (using the third option as μ is negative) as

$$T(t) = C\cos\left(\frac{cn\pi t}{L}\right) + D\sin\left(\frac{cn\pi t}{L}\right),$$

where C and D are constants.

Multiplying together the solutions for $X(x)$ and $T(t)$, and combining constants, gives the family of normal mode solutions

$$u_n(x,t) = \sin\left(\frac{n\pi x}{L}\right)\left[a_n\cos\left(\frac{cn\pi t}{L}\right) + b_n\sin\left(\frac{cn\pi t}{L}\right)\right],$$

where $a_n = BC$ and $b_n = BD$ are constants; we have added the subscript n to emphasise that there are different constants for each value of n.

The general solution of the partial differential equation with boundary conditions is a linear combination of these normal mode solutions:

$$u(x,t) = \sum_{n=1}^{\infty} \sin\left(\frac{n\pi x}{L}\right)\left[a_n\cos\left(\frac{cn\pi t}{L}\right) + b_n\sin\left(\frac{cn\pi t}{L}\right)\right].$$

Now we apply the initial conditions to find a particular solution for the plucked string model. Starting with the homogeneous initial condition, that the string starts from rest, we need the solution to satisfy $u_t(x,0) = 0$. We begin by computing the partial derivative of the general solution:

$$\frac{\partial u}{\partial t} = \sum_{n=1}^{\infty} \sin\left(\frac{n\pi x}{L}\right)\left[-\frac{cn\pi}{L}a_n\sin\left(\frac{cn\pi t}{L}\right) + \frac{cn\pi}{L}b_n\cos\left(\frac{cn\pi t}{L}\right)\right].$$

◄ Initial conditions ►

So we have

$$u_t(x,0) = \sum_{n=1}^{\infty} \sin\left(\frac{n\pi x}{L}\right)\left[\frac{cn\pi}{L}b_n\right].$$

From the results of Unit 13, if $u_t(x,0) = 0$, then by inspection $b_n = 0$ for all n. (This is a consequence of a Fourier sine series giving a unique representation of a function.)

This gives the following solution that satisfies the partial differential equation, the boundary conditions and the homogeneous initial condition:

$$u(x,t) = \sum_{n=1}^{\infty} a_n\sin\left(\frac{n\pi x}{L}\right)\cos\left(\frac{cn\pi t}{L}\right). \tag{21}$$

Now we turn to the inhomogeneous initial condition, namely $u(x,0) = f(x)$ for $0 < x < L$. Substituting for $u(x,t)$ in this equation gives

$$\sum_{n=1}^{\infty} a_n \sin\left(\frac{n\pi x}{L}\right) = f(x).$$

This is a Fourier sine series for the function $f(x)$, which is given by the odd periodic extension of the function $f(x)$ on the interval $[-L, L]$. So the coefficients a_n will be given by the formula

$$a_n = \frac{2}{2L} \int_{-L}^{L} f_{\text{odd}}(x) \sin\left(\frac{n\pi x}{L}\right) dx.$$

As the odd extension f_{odd} is odd and the sine function is odd, the product is even, so the above integral can be written as twice the sum over the positive x-values:

$$a_n = \frac{2}{L} \int_{0}^{L} f(x) \sin\left(\frac{n\pi x}{L}\right) dx,$$

where we have written $f(x)$ instead of $f_{\text{odd}}(x)$ as these two functions are equal for $0 < x < L$.

Evaluating these integrals for a given initial profile $f(x)$ will give the coefficients a_n of the solution (21) for the plucked string model.

Example 2 looked at initial conditions for a string plucked at its centre, and in this case the initial displacement $f(x)$ is given by

$$f(x) = \begin{cases} x/L & \text{for } 0 \leq x < \tfrac{1}{2}L, \\ (L-x)/L & \text{for } \tfrac{1}{2}L \leq x \leq L. \end{cases}$$

The Fourier sine series for this function was considered in Example 12 of Unit 13, where it was found that the coefficients are

$$a_n = \frac{4}{n^2 \pi^2} \sin\left(\frac{n\pi}{2}\right),$$

so the values for the coefficients a_n, for $n = 1, 2, 3, 4, 5, 6, \ldots$, are given by

$$\frac{4}{\pi^2}, \ 0, \ -\frac{4}{9\pi^2}, \ 0, \ \frac{4}{25\pi^2}, \ 0, \ \ldots .$$

Substituting into equation (21) gives the solution for our plucked string problem as

$$u(x,t) = \frac{4}{\pi^2}\left[\sin\left(\frac{\pi x}{L}\right)\cos\left(\frac{\pi ct}{L}\right) - \frac{1}{9}\sin\left(\frac{3\pi x}{L}\right)\cos\left(\frac{3\pi ct}{L}\right)\right.$$

$$\left. + \frac{1}{25}\sin\left(\frac{5\pi x}{L}\right)\cos\left(\frac{5\pi ct}{L}\right) - \cdots\right].$$

This may seem a very complicated solution, and you may not be able to see immediately how this combination of terms behaves. The behaviour is best seen graphically, as in Figure 10.

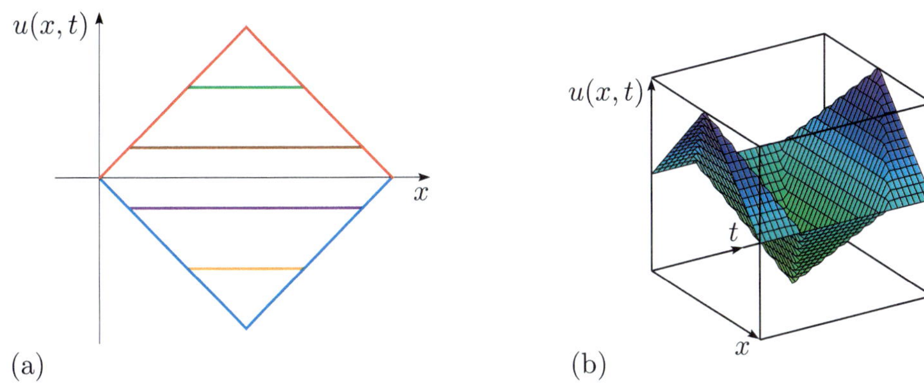

Figure 10   (a) Snapshots of the profile of the string initially (red), then after successive tenths of a cycle, green, brown, purple, orange, blue (and then back again). (b) The solution $u(x,t)$ of the damped wave equation plotted as a surface.

Figure 10 shows two different representations of the solution to the plucked string problem. Figure 10(a) shows snapshots of the position of the string at several points in time. This shows that initially only the centre portion of the string starts to move. As time progresses, more of the string moves, until after one-quarter of a cycle, the whole string is horizontal and moving downwards. Over the next quarter cycle this sequence is reversed, with the centre portion moving most, until after half a cycle the string regains its initial triangular shape. The sequence of shapes over the next half cycle is the reverse of the above, so that after one complete cycle the string has regained its initial position, then the cycle repeats. Figure 10(b) shows the same information. On the left-hand edge the string has its initial triangular shape. Moving left to right on the time axis, the string goes through one complete cycle, and on the right-hand edge we see that the string has regained its initial triangular shape. If you imagine slicing across the surface at regular time intervals, then you will see the curves shown in Figure 10(a).

This is a quite surprising prediction from the model, but modern high-speed photography confirms that this is indeed the motion of a plucked string initially. This model is only an idealised approximation of a real string, and over long time periods the approximation becomes worse. For example, real strings are not perfectly flexible and there is a force that resists the string forming a kink such as the kink at the centre of the triangular initial shape. This extra force is something that has not been modelled, and it can cause the predictions of the model to deviate from reality.

Using this model we can calculate the fundamental frequency of the guitar string that we considered in Unit 11, Section 3. There we constructed discrete models (consisting of separate springs and particles) of a guitar E string, with fundamental frequency 323 Hz, length 0.65 m and mass 0.25 g. We calculated the equilibrium tension T_{eq} of the string to be 68 N.

This model predicts that the fundamental angular frequency of the guitar string will be the smallest coefficient of t in the cosine terms of the series, that is, $\pi c/L$. The fundamental frequency f is obtained by dividing the fundamental angular frequency by 2π:

$$f = \frac{\omega}{2\pi} = \frac{\pi c/L}{2\pi} = \frac{c}{2L}.$$

Now c is related to the properties of the string by the relationship (13), namely $c^2 = T_{\text{eq}}L/M$. This can be substituted into the above equation to yield

$$f = \frac{1}{2L}\sqrt{\frac{T_{\text{eq}}L}{M}} = \frac{1}{2}\sqrt{\frac{T_{\text{eq}}}{ML}} = \frac{1}{2}\sqrt{\frac{68}{0.25 \times 10^{-3} \times 0.65}} = 323.4.$$

So this continuous model of the guitar string gives an accurate prediction of the fundamental frequency.

The following exercise looks at a different initial condition.

Exercise 14

Write down the solution for the plucked string model when the initial profile $f(x)$ is given by

$$f(x) = \frac{1}{2}\sin\left(\frac{3\pi x}{L}\right).$$

2.3 Solving the damped wave equation

In Subsection 2.2 we solved the wave equation model for vibrations of a taut string initially plucked at its midpoint. In this subsection we generalise the method in order to solve the damped wave equation model for vibrations of a damped taut string plucked at its midpoint.

The following exercise asks you to begin by following the first step of the separation of variables procedure.

Exercise 15

Consider the damped wave equation involving the constants ε and c,

$$\frac{\partial^2 u}{\partial x^2} = \frac{1}{c^2}\left(\frac{\partial^2 u}{\partial t^2} + 2\varepsilon\,\frac{\partial u}{\partial t}\right),$$

together with the boundary conditions

$$u(0,t) = u(L,t) = 0, \quad t \geq 0.$$

Apply the first step of Procedure 1 to separate the variables and obtain two ordinary differential equations together with corresponding boundary conditions.

The results of this exercise are used in the following example, which uses arguments about the relative sizes of parameters to find an approximate solution to the problem. This often happens with complicated mathematical models – either the exact equations cannot be solved or the solution is too complicated to give insight into the problem – and consequently an approximate solution to the problem is sought. This example is consequently harder than the other examples in this unit.

Example 5

Consider the model developed in Subsection 2.1 for damped vibrations of a string,

$$\frac{\partial^2 u}{\partial x^2} = \frac{1}{c^2}\left(\frac{\partial^2 u}{\partial t^2} + 2\varepsilon\,\frac{\partial u}{\partial t}\right), \tag{22}$$

where ε and c are constants. In this example we consider weak damping, which is the case when ε is small.

As in Example 4, the string is subject to fixed endpoint boundary conditions

$$u(0,t) = u(L,t) = 0, \quad t \geq 0, \tag{23}$$

and initial conditions that the string starts from rest in a given shape:

$$u(x,0) = \begin{cases} x/L & \text{for } 0 < x \leq \tfrac{1}{2}L, \\ (L-x)/L & \text{for } \tfrac{1}{2}L < x < L, \end{cases} \tag{24}$$

$$u_t(x,0) = 0, \quad 0 \leq x \leq L. \tag{25}$$

Solution

Using the results of Exercise 15, the differential equations corresponding to the given partial differential equation are   ◀ Separate variables ▶

$$X''(x) - \mu\,X(x) = 0 \tag{26}$$

and

$$T''(t) + 2\varepsilon\,T'(t) - c^2\mu\,T(t) = 0, \tag{27}$$

where μ is the separation constant. In addition, we have

$$X(0) = 0 \quad \text{and} \quad X(L) = 0, \tag{28}$$

which are boundary conditions for the ordinary differential equation (26).

The differential equation for X, and its boundary conditions, are the same   ◀ Solve ODEs ▶
as in Example 4. You have seen that a non-trivial solution occurs only if the separation constant μ is negative, and is given by

$$X(x) = A\cos kx + B\sin kx,$$

where A and B are constants, and $k = \sqrt{-\mu}$.

As in Example 4, the boundary conditions (28) tell us that $A = 0$ and $B \sin kL = 0$, so $B = 0$ or $\sin kL = 0$. The case $B = 0$ gives the trivial solution again, whereas the case $\sin kL = 0$ restricts k to take one of the values $n\pi/L$, where n is an integer. Hence, as in Example 4, we find again a family of solutions to the boundary-value problem for X of the form

$$X(x) = B \sin\left(\frac{n\pi x}{L}\right), \quad n = 1, 2, 3, \ldots . \tag{29}$$

Now we must deal with the equation for T. We know from our discussion of the function X that to have a non-trivial solution u, we must have $\mu = -k^2$ and $k = n\pi/L$ (for positive integers n), so the differential equation (27) for T becomes

$$T''(t) + 2\varepsilon\, T'(t) + \omega^2\, T(t) = 0, \quad \text{where } \omega = ck.$$

To solve this equation for T, we first solve the auxiliary equation

$$\lambda^2 + 2\varepsilon\lambda + \omega^2 = 0,$$

which gives

$$\lambda = -\varepsilon \pm \sqrt{\varepsilon^2 - \omega^2}.$$

Using the assumption that $\varepsilon \ll \omega$, that is, ε is much smaller than ω, which corresponds to very weak damping, this reduces to

$$\lambda \simeq -\varepsilon \pm i\omega.$$

The corresponding general solution is

$$T(t) \simeq e^{-\varepsilon t}(C \cos \omega t + D \sin \omega t),$$

where C and D are constants.

Using the allowed values for k (i.e. $n\pi/L$ for positive integers n), where $\omega = ck$, this gives the approximate solutions

$$T(t) \simeq e^{-\varepsilon t}\left(C \cos\left(\frac{cn\pi t}{L}\right) + D \sin\left(\frac{cn\pi t}{L}\right)\right), \quad n = 1, 2, 3, \ldots . \tag{30}$$

Now we can combine the two families (29) and (30) to obtain the family of approximate solutions

$$u_n(x, t) \simeq e^{-\varepsilon t} \sin\left(\frac{n\pi x}{L}\right)\left(a_n \cos\left(\frac{cn\pi t}{L}\right) + b_n \sin\left(\frac{cn\pi t}{L}\right)\right), \tag{31}$$
$$n = 1, 2, 3, \ldots ,$$

where $a_n = BC$ and $b_n = BD$ are constants. By the principle of superposition, any linear combination of members of this family of solutions is (approximately) a solution of the original partial differential equation (22) and satisfies the boundary conditions (23).

◀ Initial conditions ▶

Now we try to satisfy the initial conditions. We begin with the homogeneous initial condition (25). This was automatically satisfied for the undamped problem, but it is not automatically satisfied here.

We know, from the principle of superposition, that if each member of family (31) satisfies the homogeneous initial condition (25), then so will any linear combination of members of the family. So we just need to consider $u_n(x, t)$ for an arbitrary n (in the range $n = 1, 2, 3, \ldots$). Differentiating (31) with respect to t, we obtain

$$\frac{\partial u_n}{\partial t} \simeq e^{-\varepsilon t} \sin\left(\frac{n\pi x}{L}\right) \left(-a_n \frac{cn\pi}{L} \sin\left(\frac{cn\pi t}{L}\right) + b_n \frac{cn\pi}{L} \cos\left(\frac{cn\pi t}{L}\right)\right)$$
$$- \varepsilon e^{-\varepsilon t} \sin\left(\frac{n\pi x}{L}\right)\left(a_n \cos\left(\frac{cn\pi t}{L}\right) + b_n \sin\left(\frac{cn\pi t}{L}\right)\right).$$

When $t = 0$, and writing $\omega = ck = cn\pi/L$ as earlier, we obtain

$$\left.\frac{\partial u_n}{\partial t}\right|_{t=0} \simeq \left(b_n \frac{cn\pi}{L} - \varepsilon a_n\right) \sin\left(\frac{n\pi x}{L}\right) = (\omega b_n - \varepsilon a_n) \sin\left(\frac{n\pi x}{L}\right).$$

For initial condition (25) to be satisfied, for $0 \leq x \leq L$, we must have $\omega b_n - \varepsilon a_n \simeq 0$. For very weak damping, as we noted earlier, we have $\varepsilon \ll \omega$. This means that to satisfy $\omega b_n - \varepsilon a_n \simeq 0$, we must have $b_n \simeq 0$. Thus to satisfy the initial condition (25), approximately, we need to restrict family (31) to

$$u_n(x, t) \simeq a_n e^{-\varepsilon t} \sin\left(\frac{n\pi x}{L}\right) \cos\left(\frac{cn\pi t}{L}\right).$$

We now need to satisfy the inhomogeneous initial condition (24). As in Procedure 1, we look for an approximate solution of the form

$$u(x, t) \simeq e^{-\varepsilon t} \sum_{n=1}^{\infty} a_n \sin\left(\frac{n\pi x}{L}\right) \cos\left(\frac{cn\pi t}{L}\right). \tag{32}$$

Setting $t = 0$ in approximation (32) gives

$$u(x, 0) \simeq \sum_{n=1}^{\infty} a_n \sin\left(\frac{n\pi x}{L}\right), \tag{33}$$

while the given initial condition (24) is

$$u(x, 0) = \begin{cases} x/L & \text{for } 0 < x \leq \tfrac{1}{2}L, \\ (L-x)/L & \text{for } \tfrac{1}{2}L < x < L. \end{cases}$$

The situation is exactly the same as in Subsection 2.2, so we obtain the same Fourier coefficients a_n as there. Substituting these into approximation (32) gives the final form of the approximate solution as

$$u(x, t) \simeq \frac{4}{\pi^2} e^{-\varepsilon t} \left[\sin\left(\frac{\pi x}{L}\right) \cos\left(\frac{\pi c t}{L}\right) - \frac{1}{9} \sin\left(\frac{3\pi x}{L}\right) \cos\left(\frac{3\pi c t}{L}\right)\right.$$
$$\left. + \frac{1}{25} \sin\left(\frac{5\pi x}{L}\right) \cos\left(\frac{5\pi c t}{L}\right) - \cdots \right].$$

Remember that this is an approximate solution, under the assumption of very weak damping, where ε is small compared with ω.

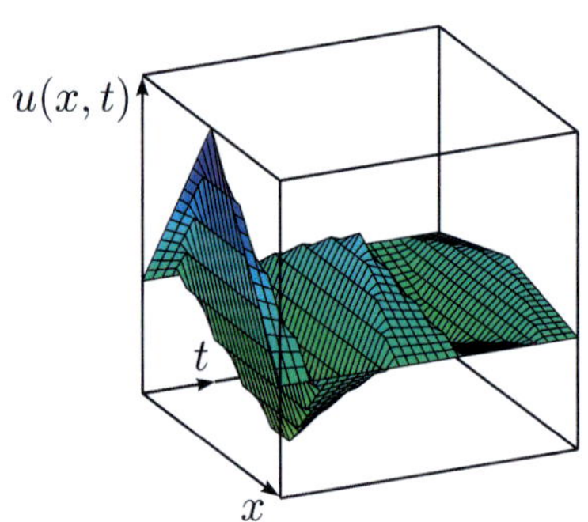

Figure 11   The solution $u(x,t)$ of the damped wave equation shown over two complete cycles

This is the same as the solution to the undamped problem except for the factor $e^{-\varepsilon t}$, which describes exponential decay. Hence the shape of the string is roughly the same as before, apart from the amplitude becoming progressively smaller by the factor $e^{-\varepsilon t}$. The shape determines the sound, so the sound stays the same, only now it gets progressively quieter, since the volume is determined by the amplitude. The solution is shown in Figure 11.

The figure shows that the exponential factor has the effect of reducing the amplitude of the vibrations of the string, which is to be expected for a solution of the damped wave equation.

3 The heat equation

In this section we look at a situation that can be modelled using a different second-order partial differential equation, known as the heat equation. In Subsection 3.1 we will see how the heat equation arises as a model of temperature variation in an insulated metal rod, that is, a rod that does not exchange heat with its surroundings. In addition, we will see how, for an uninsulated metal rod, heat exchange with its surroundings can be incorporated into the model.

Then in Subsection 3.2 we will see how the method of separating the variables, Procedure 1, can be used to solve the heat equation subject to appropriate boundary and initial conditions.

3.1 Deriving the heat equation

This derivation is given for completeness – you will not be asked to reproduce or modify this derivation in the assessment of this module.

In order to model the temperature changes in a rod, we need a formula to describe how temperature changes. Basic intuition tells us that hot things cool down and cold things heat up. Refining this a little, we might consider how we distinguish between hot and cold, and a moment's thought will reveal that hot and cold are relative to the temperature of the surrounding environment. Moreover, the hotter an object is, the quicker its temperature will decrease. Isaac Newton formulated this as an empirical law of temperature change.

Newton's law of cooling

For a given object, the rate of decrease of temperature is proportional to the excess temperature over the environment.

This law applies to objects that are not changing state in the process (such as a solid melting into a liquid). Although Newton called this a law of nature, it is really just a first model of how heat flows. It should be considered in a similar way to Hooke's law for springs, where we used the law to derive useful models of oscillating systems.

We will use Θ to denote the temperature, so this law can be formulated as

$$\frac{d\Theta}{dt} = -k(\Theta - \Theta_0),$$

where Θ_0 is the temperature of the surrounding environment, and k is a positive constant of proportionality. The negative sign in this equation is because hotter objects (i.e. with $\Theta > \Theta_0$) cool down (so $d\Theta/dt < 0$). The constant of proportionality k depends on many properties of the object (such as size and mass) and the contact between the object and its surroundings (such as whether there is direct contact or an air gap), but will be constant for a given object in a given situation.

The study of the flow of heat is called thermodynamics, and this is a substantial topic in modern physics. Here we are not concerned with the details of thermodynamics and the different methods of heat transfer. We will use Newton's law of cooling as a mathematical model of how everyday physical objects behave in an analogous way to how we use Hooke's law as a mathematical model to describe the behaviour of physical springs.

Armed with this mathematical model, we can derive a partial differential equation that models the change in temperature distribution in a uniform rod as it conducts heat. We will assume that no heat is lost from the sides of the rod and that heat is transferred only along the length of the rod (so that we have a one-dimensional problem). We are modelling a straight rod, and we choose an x-axis to be aligned with the rod with the origin at the left-hand end. Let $\Theta(x,t)$ be the temperature at position x along the rod at time t. Consider a small segment of rod of length δx with midpoint a distance x along the rod, as shown in Figure 12.

We will now apply our model of cooling with the small central segment of the rod considered as the object. The temperature on the left of the central segment is $\Theta(x - \delta x, t)$ and this takes the role of the temperature of the environment for the interface between the central and left-hand segments, so the temperature change due to this left-hand interface is $-k\big(\Theta(x,t) - \Theta(x - \delta x, t)\big)$. Similarly, on the right-hand end of the central segment, the 'environment' temperature is $\Theta(x + \delta x, t)$, and the temperature change due to heat exchange at this interface is $-k\big(\Theta(x,t) - \Theta(x + \delta x, t)\big)$ (note that the constant of proportionality k is the same for both ends as the rod is uniform thus the ends are identical). The change in temperature of the central segment $\Theta_t(x,t)$ is given by the sum of these two contributions:

$$\Theta_t(x,t) = -k\big(\Theta(x,t) - \Theta(x - \delta x, t)\big) - k\big(\Theta(x,t) - \Theta(x + \delta x, t)\big)$$
$$= k\big(\Theta(x - \delta x, t) - 2\Theta(x,t) + \Theta(x + \delta x, t)\big). \tag{34}$$

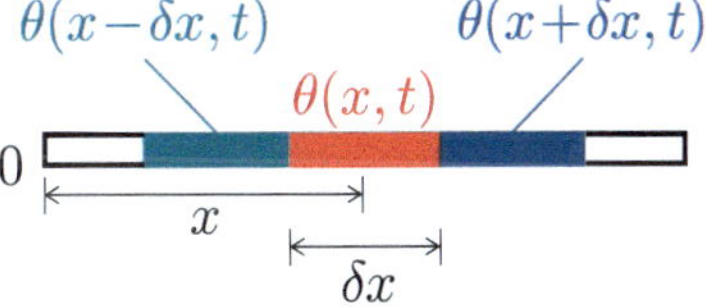

Figure 12   A conducting rod

This can be simplified by using the following Taylor series:

$$\Theta(x + \delta x, t) = \Theta(x,t) + \delta x\,\Theta_x(x,t) + \tfrac{1}{2}(\delta x)^2\,\Theta_{xx}(x,t) + \cdots,$$
$$\Theta(x - \delta x, t) = \Theta(x,t) - \delta x\,\Theta_x(x,t) + \tfrac{1}{2}(\delta x)^2\,\Theta_{xx}(x,t) - \cdots.$$

Adding these expressions gives

$$\Theta(x - \delta x, t) - 2\Theta(x,t) + \Theta(x + \delta x, t) = (\delta x)^2\,\Theta_{xx}(x,t) + \cdots,$$

so

$$\Theta_t(x,t) \simeq k\,(\delta x)^2\,\Theta_{xx}(x,t).$$

As δx becomes smaller, the Taylor approximation will become exact and we can replace the above approximation with an equality. Also, the constant $k\,(\delta x)^2$ must tend to a finite limit α so that the model of smoothly varying temperature applies. (If $k\,(\delta x)^2$ tends to infinity, then $\partial\Theta/\partial t$ tends to infinity, that is, the temperature jumps abruptly.) This gives the partial differential equation model that we are aiming to derive.

Note that k varies with δx, as it is the constant of proportionality for a segment of the rod of length δx.

Heat equation

$$\frac{\partial\Theta}{\partial t} = \alpha\,\frac{\partial^2\Theta}{\partial x^2}. \tag{35}$$

If the ends of the rod are kept at a steady temperature Θ_0, the boundary conditions are

$$\Theta(0,t) = \Theta(L,t) = \Theta_0, \quad t \geq 0. \tag{36}$$

Further, if the initial distribution of temperature along the rod is given by a function $f(x)$, then we have the initial condition

Note that we need only one initial condition because the equation is first-order in time.

$$\Theta(x,0) = f(x), \quad 0 < x < L. \tag{37}$$

These conditions are sufficient for us to be able to obtain a unique solution of the heat equation for the variation of temperature in an insulated rod, as you will see in Subsection 3.2.

Exercise 16

Show that the function

$$\Theta(x,t) = \sin\left(\frac{\pi x}{L}\right) \exp\left(-\frac{\alpha\pi^2 t}{L^2}\right)$$

satisfies the heat equation (35) and the boundary conditions (36) if $\Theta_0 = 0$.

Exercise 17

Suppose that initially, the temperature of a rod rises linearly with x towards a peak in the centre that is half a degree above the end temperature Θ_0, as shown in Figure 13.

Write down a formula describing the initial temperature function $f(x)$.

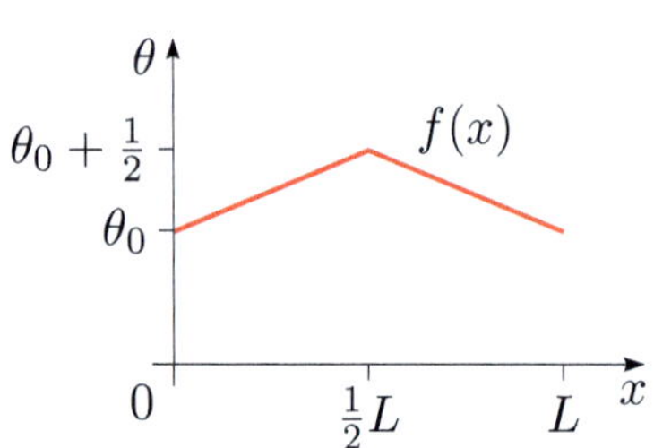

Figure 13 An initial temperature distribution

Exercise 18

Write down the initial condition describing the temperature distribution if the central third of a rod is initially heated to a temperature Θ_1 while the remainder of the rod stays at the background temperature Θ_0.

We conclude this subsection by considering a rod that is not insulated from the environment, so that the rod loses heat to an environment of constant temperature Θ_0 in addition to conducting heat along its length. In this situation, equation (34) must be modified to include a term that represents the heat transferred to its surroundings along the length of the segment in addition to the heat transferred at the ends. Applying our model of cooling, this term will be $-\gamma(\Theta - \Theta_0)$, where Θ_0 is the temperature of the surroundings, and we have written the constant of proportionality as γ to distinguish it from the different constant of proportionality k already in the equation. So we have

$$\Theta_t(x,t) = k\big(\Theta(x - \delta x, t) - 2\Theta(x,t) + \Theta(x + \delta x, t)\big) - \gamma(\Theta - \Theta_0),$$

and proceeding by using Taylor series as before, we can derive the following partial differential equation.

Uninsulated rod equation

$$\frac{\partial \Theta}{\partial t} = \alpha \frac{\partial^2 \Theta}{\partial x^2} - \gamma(\Theta - \Theta_0). \tag{38}$$

3.2 Solving the heat equation

In this subsection we will use Procedure 1 to solve a particular case of the heat equation that was derived in Subsection 3.1. In Example 6 we consider a rod insulated from its surroundings with a given initial temperature.

Example 6

Apply Procedure 1 to the heat equation

$$\frac{\partial \Theta}{\partial t} = \alpha \frac{\partial^2 \Theta}{\partial x^2},$$

subject to the boundary conditions

$$\Theta_x(0, t) = \Theta_x(L, t) = 0, \quad t \geq 0,$$

These boundary conditions model the situation where the ends of a hot rod are insulated.

and the initial condition

$$\Theta(x, 0) = \tfrac{1}{2} \cos\left(\frac{2\pi x}{L}\right), \quad 0 < x < L.$$

Solution

We write $\Theta(x, t) = X(x)\, T(t)$, so

$$\frac{\partial \Theta}{\partial x} = X'T, \quad \frac{\partial^2 \Theta}{\partial x^2} = X''T \quad \text{and} \quad \frac{\partial \Theta}{\partial t} = XT'.$$

Substituting into the partial differential equation gives $XT' = \alpha X''T$. Separate the variables to yield

$$\frac{X''}{X} = \frac{T'}{\alpha T}.$$

Arguing as before, both sides of this equation must be equal to a separation constant μ, so we have the two differential equations

$$X'' - \mu X = 0 \quad \text{and} \quad T' - \alpha\mu T = 0.$$

We have $\partial \Theta / \partial x = X'T$, so the boundary conditions become

$$X'(0)\, T(t) = X'(L)\, T(t) = 0, \quad t \geq 0,$$

hence

$$X'(0) = X'(L) = 0.$$

Consider the three cases $\mu = k^2 > 0$, $\mu = 0$ and $\mu = -k^2 < 0$.

- $\mu > 0$. Let $c = \sqrt{\mu}$. Then equation (9) gives the solution of the differential equation for $X(x)$ as

  $$X(x) = Ae^{cx} + Be^{-cx},$$

  where A and B are constants.

  Now we apply the boundary conditions, and to do this we first differentiate this solution:

  $$X'(x) = Ace^{cx} - Bce^{-cx}.$$

  The boundary condition $X'(0) = 0$ gives $Ac - Bc = 0$, so $A = B$ as $c > 0$. The boundary condition $X'(L) = 0$ gives

  $$Ace^{cL} - Bce^{-cL} = 0.$$

  As $A = B$ and $c > 0$, this simplifies to

  $$A(e^{cL} - e^{-cL}) = 0.$$

  Proceeding as before we can multiply by e^{cL} to obtain

  $$A(e^{2cL} - 1) = 0.$$

  Now we can see that the term in brackets is not zero as $2cL > 0$ so $\exp(2cL) > \exp(0) = 1$. So we have $A = B = 0$, and the only solution is the trivial solution.

- $\mu = 0$. In this case the differential equation for $X(x)$ becomes $X'' = 0$, with solution $X(x) = Ax + B$, where A and B are constants. So $X'(x) = A$, and both boundary conditions give the equation $A = 0$.

  So the solution $X(x) = B$ is a non-trivial solution of the equation.

- $\mu < 0$. Let $k = \sqrt{-\mu}$. Then by equation (9) the general solution of the equation for X is

$$X(x) = A\cos kx + B\sin kx,$$

where A and B are constants, so

$$X'(x) = -Ak\sin kx + Bk\cos kx.$$

Using the boundary conditions, we find that $B = 0$, and $A = 0$ or $k = n\pi/L$ for any non-zero integer n. This leads to the solution

$$X(x) = A\cos\left(\frac{n\pi x}{L}\right), \quad n = 1, 2, 3, \ldots.$$

So we have *two* cases that give non-trivial solutions, namely the constant solutions that correspond to $\mu = 0$ and the sinusoidal solutions that correspond to $\mu < 0$. Here these two cases can be conveniently combined by adding $n = 0$ to the second set of solutions (using the fact that $\cos(0 \times x) = 1$ for all x). So we have the solutions

$$X(x) = A\cos\left(\frac{n\pi x}{L}\right), \quad n = 0, 1, 2, 3, \ldots.$$

With $\mu = n^2\pi^2/L^2$, the differential equation for $T(t)$ becomes

$$T'(t) + \alpha\,\frac{n^2\pi^2}{L^2}\,T(t) = 0.$$

Equation (8) gives the solution of this equation as

$$T(t) = C\exp\left(-\frac{\alpha n^2\pi^2 t}{L^2}\right), \quad n = 0, 1, 2, 3, \ldots,$$

where C is a constant.

This leads to the family of solutions

$$\Theta_n(x,t) = a_n\exp\left(-\frac{\alpha n^2\pi^2 t}{L^2}\right)\cos\left(\frac{n\pi x}{L}\right), \quad n = 0, 1, 2, 3, \ldots,$$

where the $a_n = AC$ are constants.

To find the solution that satisfies the initial condition, write

$$\Theta(x,t) = \sum_{n=0}^{\infty} a_n\exp\left(-\frac{\alpha n^2\pi^2 t}{L^2}\right)\cos\left(\frac{n\pi x}{L}\right), \tag{39}$$

◀ Initial conditions ▶

and set $t = 0$ to give

$$\Theta(x,0) = \sum_{n=0}^{\infty} a_n\cos\left(\frac{n\pi x}{L}\right).$$

Now we use the given initial condition

$$\Theta(x,0) = \tfrac{1}{2}\cos\left(\frac{2\pi x}{L}\right).$$

By inspection we have $a_2 = \tfrac{1}{2}$, and $a_n = 0$ for $n \neq 2$, so the solution is

$$\Theta(x,t) = \tfrac{1}{2}\exp\left(-\frac{4\alpha\pi^2 t}{L^2}\right)\cos\left(\frac{2\pi x}{L}\right).$$

In Example 6, if the initial condition was not a cosine function, then at this point we would need to use the techniques of Unit 13 to find the Fourier cosine series in order to find a_n. Recall that the Fourier coefficient a_0 is equal to the average value of the given function over the interval. In this case this would mean that a_0 would be the average of the initial temperature distribution over the length of the rod. Looking at equation (39), we can see that the term involving a_0 is the only term that does not involve a negative exponential function of time. So the model predicts that the temperature of the rod tends to a constant temperature that is equal to the average of the initial temperatures. This is in accord with intuition for a model of a rod that is completely insulated from its surroundings.

The following exercises ask you to apply the separation of variables method.

Exercise 19

Solve the heat equation

$$\frac{\partial \Theta}{\partial t} = \alpha \frac{\partial^2 \Theta}{\partial x^2}, \tag{40}$$

These boundary conditions model the situation where the ends of the rod are held at a fixed temperature.

subject to the boundary conditions

$$\Theta(0, t) = \Theta(L, t) = 0, \quad t \geq 0, \tag{41}$$

and the initial condition

$$\Theta(x, 0) = \sin\left(\frac{\pi x}{L}\right). \tag{42}$$

Exercise 20

Find the solution of the uninsulated rod equation

This is equation (38) derived in Subsection 3.1, with $\Theta_0 = 0$.

$$\frac{\partial \Theta}{\partial t} = \alpha \frac{\partial^2 \Theta}{\partial x^2} - \gamma \Theta, \tag{43}$$

subject to the boundary conditions

$$\Theta(0, t) = \Theta(L, t) = 0, \quad t \geq 0, \tag{44}$$

and the initial condition

$$\Theta(x, 0) = \sin\left(\frac{\pi x}{L}\right). \tag{45}$$

Compare the solution with the solution to the insulated rod problem in Example 6.

Learning outcomes

After studying this unit, you should be able to:

- use the terms linear, homogeneous, order, initial condition and boundary condition as applied to partial differential equations
- show that a given function satisfies a given partial differential equation and/or boundary conditions and/or initial conditions
- use the method of separation of variables to find solutions of linear homogeneous second-order partial differential equations
- understand how the wave equation and heat equation can be used to model certain physical systems
- interpret solutions of partial differential equations in terms of a model.

Solutions to exercises

Solution to Exercise 1

(a) This equation is non-linear (because of the term that contains a product of u and u_x). It is a first-order equation.

(b) This equation is a linear second-order equation.

(c) This equation is a linear second-order equation.

(d) This equation is a linear third-order equation.

Solution to Exercise 2

Start by differentiating $u(x,t)$:

$$\frac{\partial u}{\partial x} = \cos(x)\, e^{-\alpha t}, \quad \frac{\partial^2 u}{\partial x^2} = -\sin(x)\, e^{-\alpha t}, \quad \frac{\partial u}{\partial t} = -\alpha \sin(x)\, e^{-\alpha t}.$$

So

$$\alpha \frac{\partial^2 u}{\partial x^2} = -\alpha \sin(x)\, e^{-\alpha t} = \frac{\partial u}{\partial t},$$

and the partial differential equation is satisfied.

Solution to Exercise 3

The initial velocity is zero, so the string is released from rest. When $x = \frac{1}{4}L$, $u(x,0) = -d$, so the point one-quarter of the way along the string has been displaced downwards by a distance d.

Solution to Exercise 4

The initial displacement has two linear sections, with slopes

$$\frac{d}{\frac{1}{3}L} = \frac{3d}{L} \quad \text{and} \quad -\frac{d}{\frac{2}{3}L} = -\frac{3d}{2L},$$

respectively. Hence the required initial condition is

$$u(x,0) = \begin{cases} \dfrac{3d}{L}x & \text{for } 0 < x \le \tfrac{1}{3}L, \\[2mm] \dfrac{3d}{2L}(L-x) & \text{for } \tfrac{1}{3}L < x < L. \end{cases}$$

Solution to Exercise 5

Initially, there is no displacement, so

$$u(x,0) = 0, \quad 0 < x < L.$$

Since only the middle third is set in motion, the initial velocity is given by

$$u_t(x,0) = \begin{cases} v & \text{for } \tfrac{1}{3}L \le x \le \tfrac{2}{3}L, \\ 0 & \text{otherwise.} \end{cases}$$

Solution to Exercise 6

The boundary conditions are satisfied since

$$u(0,t) = \sin 0 \cos\left(\frac{\pi c t}{L}\right) = 0,$$

$$u(L,t) = \sin \pi \cos\left(\frac{\pi c t}{L}\right) = 0.$$

The initial condition is satisfied since

$$u_t(x,0) = -\frac{\pi c}{L}\sin\left(\frac{\pi x}{L}\right)\sin 0 = 0.$$

Solution to Exercise 7

(a) Since the function $\Theta(x,t) = \Theta_0$ is constant, all its derivatives are zero, and the differential equation reduces to $0 = 0$ thus is satisfied.

As $\Theta(x,t)$ is always equal to Θ_0, in particular it is equal to Θ_0 at the boundaries, that is, the boundary conditions are satisfied.

(b) As Θ_0 is constant, its derivatives are zero, so $u_t = \Theta_t$ and $u_{xx} = \Theta_{xx}$. So if Θ satisfies the heat equation, then so does u.

On the boundary we have $\Theta(0,t) = \Theta_0$, so $u(0,t) = \Theta_0 - \Theta_0 = 0$. Similarly, $\Theta(L,t) = \Theta_0$, so $u(L,t) = \Theta_0 - \Theta_0 = 0$.

So $u(x,t)$ satisfies the heat equation with homogeneous boundary conditions.

Solution to Exercise 8

(a) For

$$\Theta(x,t) = \frac{L-x}{L}\Theta_0 + \frac{x}{L}\Theta_L,$$

we have

$$\frac{\partial^2 \Theta}{\partial x^2} = 0 \quad \text{and} \quad \frac{\partial \Theta}{\partial t} = 0,$$

so the differential equation reduces to $0 = 0$ thus is satisfied.

When $x = 0$ we have

$$\Theta(0,t) = \frac{L-0}{L}\Theta_0 + 0 = \Theta_0,$$

and when $x = L$ we have

$$\Theta(L,t) = \frac{L-L}{L}\Theta_0 + \frac{L}{L}\Theta_L = \Theta_L,$$

so the boundary conditions are satisfied.

(b) Start by differentiating:

$$\frac{\partial u}{\partial t} = \frac{\partial \Theta}{\partial t},$$

$$\frac{\partial u}{\partial x} = \frac{\partial \Theta}{\partial x} + \frac{1}{L}\Theta_0 - \frac{1}{L}\Theta_L,$$

$$\frac{\partial^2 u}{\partial x^2} = \frac{\partial^2 \Theta}{\partial x^2}.$$

So $u(x,t)$ satisfies the heat equation if $\Theta(x,t)$ satisfies the heat equation.

For the first boundary condition we have

$$u(0,t) = \Theta(0,t) - \frac{L-0}{L}\Theta_0 - 0 = \Theta_0 - \Theta_0 = 0,$$

and for the second boundary condition we have

$$u(L,t) = \Theta(L,t) - \frac{L-L}{L}\Theta_0 - \frac{L}{L}\Theta_L = \Theta_L - \Theta_L = 0.$$

So $u(x,t)$ satisfies the heat equation with homogeneous boundary conditions.

Solution to Exercise 9

$$\frac{\partial u}{\partial t} = X(x)\,T'(t), \quad \frac{\partial^2 u}{\partial t^2} = X(x)\,T''(t),$$

$$\frac{\partial u}{\partial x} = X'(x)\,T(t), \quad \frac{\partial^2 u}{\partial x^2} = X''(x)\,T(t).$$

Solution to Exercise 10

Proceeding as in the main text, we use the derivatives found in Exercise 9, namely

$$\frac{\partial^2 u}{\partial x^2} = X''(x)\,T(t) \quad \text{and} \quad \frac{\partial u}{\partial t} = X(x)\,T'(t).$$

Substituting these into the partial differential equation gives

$$X''(x)\,T(t) + 2\,X(x)\,T(t) = X(x)\,T'(t).$$

Dividing by $X(x)\,T(t)$ gives

$$\frac{X''(x)}{X(x)} + 2 = \frac{T'(t)}{T(t)}.$$

We chose to leave the 2 on the left-hand side. It would be equally correct to write the separated equations as

$$\frac{X''(x)}{X(x)} = \frac{T'(t)}{T(t)} - 2.$$

This has separated the variables, so both sides must be equal to a constant, say μ. So we have the two equations

$$\frac{X''(x)}{X(x)} + 2 = \mu \quad \text{and} \quad \frac{T'(t)}{T(t)} = \mu.$$

Multiplying out the fractions and rearranging gives the required differential equations:

$$X''(x) + (2-\mu)\,X(x) = 0 \quad \text{and} \quad T'(t) - \mu\,T(t) = 0.$$

Solution to Exercise 11

As the boundary condition is defined by the partial derivative with respect to x, we first find

$$\frac{\partial u}{\partial x} = X'(x)\, T(t).$$

So the boundary condition becomes $X'(1)\, T(t) = 0$ for $t \geq 0$. The solution $T(t) = 0$ for $t \geq 0$ leads to the trivial solution. So for a non-trivial solution we must have $X'(1) = 0$.

Solution to Exercise 12

Setting $u(x,t) = X(x)\, T(t)$, the required partial derivatives are

◀ Separate variables ▶

$$\frac{\partial^2 u}{\partial x^2} = X''T \quad \text{and} \quad \frac{\partial^2 u}{\partial t^2} = XT''.$$

Substituting into the partial differential equation and dividing by XT gives

$$\frac{X''}{X} + \frac{T''}{T} = 0,$$

from which it follows that

$$\frac{X''}{X} = -\frac{T''}{T}.$$

Both sides of this equation must be a constant, say μ, giving

$$\frac{X''}{X} = \mu \quad \text{and} \quad -\frac{T''}{T} = \mu$$

or equivalently,

$$X'' - \mu X = 0 \quad \text{and} \quad T'' + \mu T = 0.$$

The boundary conditions become $X(0) = X(1) = 0$.

The boundary conditions are the same as those used in the main text. So arguing as in the text, only negative μ gives a non-trivial solution for X. In this case, the equation for X has general solution (see equation (9))

◀ Solve ODEs ▶

$$X(x) = A \cos kx + B \sin kx,$$

where $k = \sqrt{-\mu}$, and A and B are constants. The boundary condition $X(0) = 0$ implies $A = 0$. The boundary condition $X(1) = 0$ implies $B \sin k = 0$, so for non-trivial solutions we must have $\sin k = 0$, that is, $k = n\pi$ for some positive integer n. For this value of k we have

$$X(x) = B \sin(n\pi x), \quad n = 1, 2, 3, \ldots.$$

Since $\mu = -n^2\pi^2$, the equation for T can be written as

$$T''(t) - n^2\pi^2\, T(t) = 0.$$

Using equation (9), the solutions in this case are (as $n^2\pi^2$ is positive)

$$T(t) = Ce^{n\pi t} + De^{-n\pi t}, \quad n = 1, 2, 3, \ldots,$$

where C and D are constants.

So the required family of normal mode solutions is

$$u_n(x,t) = \sin(n\pi x)\left(a_n e^{n\pi t} + b_n e^{-n\pi t}\right),$$

where n is a positive integer and $a_n = BC$, $b_n = BD$ are constants.

Solution to Exercise 13

T_{eq} is a component of a force, so $[T_{\text{eq}}] = \mathrm{M\,L\,T^{-2}}$. Also, $[M] = \mathrm{M}$ and $[L] = \mathrm{L}$. Hence

$$[T_{\text{eq}}L/M] = \mathrm{M\,L\,T^{-2}} \times \mathrm{L/M} = \mathrm{L^2\,T^{-2}},$$

so $[\sqrt{T_{\text{eq}}L/M}] = \mathrm{L\,T^{-1}}$, as required.

Solution to Exercise 14

To find the solution we need to find the coefficient a_n appearing in the equation

$$\sum_{n=1}^{\infty} a_n \sin\left(\frac{n\pi x}{L}\right) = \frac{1}{2}\sin\left(\frac{3\pi x}{L}\right).$$

This can be done by inspection as the term on the right-hand side appears as one of the terms on the left-hand side. So we get $a_3 = \frac{1}{2}$, and $a_n = 0$ for $n \neq 3$.

So the solution for the model is given by equation (21) with the above coefficients, that is,

$$u(x,t) = \frac{1}{2}\sin\left(\frac{3\pi x}{L}\right)\cos\left(\frac{3c\pi t}{L}\right).$$

Solution to Exercise 15

Write the unknown function as

$$u(x,t) = X(x)\,T(t).$$

Differentiate to get the partial derivatives:

$$\frac{\partial u}{\partial t} = X(x)\,T'(t), \quad \frac{\partial^2 u}{\partial t^2} = X(x)\,T''(t), \quad \frac{\partial^2 u}{\partial x^2} = X''(x)\,T(t).$$

Substituting these into the damped wave equation gives

$$X''(x)\,T(t) = \frac{1}{c^2}\left(X(x)\,T''(t) + 2\varepsilon\,X(x)\,T'(t)\right).$$

As before, this equation can hold for all x and all t only if both sides are equal to a separation constant μ, that is,

$$\frac{X''(x)}{X(x)} = \frac{1}{c^2}\left(\frac{T''(t)}{T(t)} + 2\varepsilon\,\frac{T'(t)}{T(t)}\right) = \mu,$$

giving a pair of equations.

The first of these is

$$\frac{X''(x)}{X(x)} = \mu,$$

which on rearranging becomes

$$X''(x) - \mu\,X(x) = 0.$$

Similarly, the second equation reduces to

$$T''(t) + 2\varepsilon\,T'(t) - c^2\mu\,T(t) = 0.$$

Now we have completed the process of separating the single partial differential equation into two ordinary differential equations, but we still have to find boundary conditions for the function X. To do this, we substitute $u(x,t) = X(x)\,T(t)$ into the original boundary conditions.

The boundary condition $u(0,t) = 0$ becomes $X(0)\,T(t) = 0$ for all t. So for non-trivial solutions we must have $X(0) = 0$. Similarly, the boundary condition $u(L,t) = 0$ gives $X(L)\,T(t) = 0$ for all t, so non-trivial solutions must satisfy $X(L) = 0$. So the boundary conditions become

$$X(0) = 0 \quad \text{and} \quad X(L) = 0,$$

which are boundary conditions for the ordinary differential equation for X.

Solution to Exercise 16

$$\frac{\partial \Theta}{\partial x} = \frac{\pi}{L}\cos\left(\frac{\pi x}{L}\right)\exp\left(-\frac{\alpha \pi^2 t}{L^2}\right),$$

so

$$\frac{\partial^2 \Theta}{\partial x^2} = -\frac{\pi^2}{L^2}\sin\left(\frac{\pi x}{L}\right)\exp\left(-\frac{\alpha \pi^2 t}{L^2}\right) = -\frac{\pi^2}{L^2}\Theta(x,t).$$

Also,

$$\frac{\partial \Theta}{\partial t} = -\frac{\alpha \pi^2}{L^2}\sin\left(\frac{\pi x}{L}\right)\exp\left(-\frac{\alpha \pi^2 t}{L^2}\right) = -\frac{\alpha \pi^2}{L^2}\Theta(x,t).$$

Thus

$$\frac{\partial \Theta}{\partial t} = -\alpha\,\frac{\pi^2}{L^2}\Theta(x,t) = \alpha\,\frac{\partial^2 \Theta}{\partial x^2}.$$

Hence equation (35) is satisfied.

The boundary conditions are

$$\Theta(0,t) = \exp\left(-\frac{\alpha \pi^2 t}{L^2}\right)\sin 0 = 0,$$

$$\Theta(L,t) = \exp\left(-\frac{\alpha \pi^2 t}{L^2}\right)\sin \pi = 0,$$

so these are satisfied for a temperature at the rod ends of $\Theta_0 = 0$.

Solution to Exercise 17

The graph in Figure 13 looks just like the picture of the plucked string in Figure 2, so the required formula is

$$
f(x) = \begin{cases} \Theta_0 + x/L & \text{for } 0 < x \le \tfrac{1}{2}L, \\ \Theta_0 + (L-x)/L & \text{for } \tfrac{1}{2}L < x < L. \end{cases}
$$

As required, this function takes the value Θ_0 when x is 0 or L, and takes the value $\Theta_0 + \tfrac{1}{2}$ when x is $\tfrac{1}{2}L$.

Solution to Exercise 18

$$
\Theta(x,0) = \begin{cases} \Theta_0 & \text{for } 0 < x < \tfrac{1}{3}L, \\ \Theta_1 & \text{for } \tfrac{1}{3}L \le x \le \tfrac{2}{3}L, \\ \Theta_0 & \text{for } \tfrac{2}{3}L < x < L. \end{cases}
$$

Solution to Exercise 19

◀ Separate variables ▶

We begin as in Example 6, and obtain the equations

$$
X'' - \mu X = 0 \quad \text{and} \quad T' - \alpha\mu T = 0.
$$

To find boundary conditions for X, we put $x = 0$ and $x = L$ in $\Theta(x,t) = X(x)\,T(t)$ and substitute into the boundary conditions (41), which gives

$$
X(0)\,T(t) = X(L)\,T(t) = 0, \quad t \ge 0,
$$

hence

$$
X(0) = X(L) = 0.
$$

◀ Solve ODEs ▶

Next we solve the differential equations for X and T, and combine the families of solutions.

The differential equation for X, and its boundary conditions, are the same as in Example 4. You have seen that a non-trivial solution occurs only if the separation constant μ is negative, and is given by

$$
X(x) = B \sin\left(\frac{n\pi x}{L}\right), \quad n = 1, 2, 3, \ldots,
$$

where B is a constant. In this case the separation constant μ is $-n^2\pi^2/L^2$.

Our next task is to solve the equation for T when $\mu = -n^2\pi^2/L^2$, namely

$$
T'(t) + \frac{\alpha n^2 \pi^2}{L^2}\,T(t) = 0.
$$

The general solution is given by equation (8) as

$$
T(t) = C \exp\left(-\frac{\alpha n^2 \pi^2 t}{L^2}\right),
$$

where C is a constant.

Combining the solutions for X and T, we obtain the family of normal mode solutions

$$\Theta_n(x,t) = a_n \exp\left(-\frac{\alpha n^2 \pi^2 t}{L^2}\right) \sin\left(\frac{n\pi x}{L}\right), \quad n = 1, 2, 3, \ldots,$$

where the $a_n = BC$ are constants.

As the partial differential equation and boundary conditions are linear and homogeneous, we can use the principle of superposition to form the more general solution

$$\Theta(x,t) = \sum_{n=1}^{\infty} a_n \exp\left(-\frac{\alpha n^2 \pi^2 t}{L^2}\right) \sin\left(\frac{n\pi x}{L}\right).$$

Now we use the initial condition (42) and results on Fourier series to determine the coefficients a_n. Setting $t = 0$ in this equation gives ◀ Initial conditions ▶

$$\sin\left(\frac{\pi x}{L}\right) = \Theta(x,0) = \sum_{n=1}^{\infty} a_n \sin\left(\frac{n\pi x}{L}\right).$$

By inspection, $a_1 = 1$, and $a_n = 0$ for $n = 2, 3, \ldots$, as the term on the left-hand side is one member of the sum on the right-hand side.

So the solution is

$$\Theta(x,t) = \exp\left(-\frac{\alpha \pi^2 t}{L^2}\right) \sin\left(\frac{\pi x}{L}\right).$$

Solution to Exercise 20

Set $\Theta(x,t) = X(x)\,T(t)$. Then ◀ Separate variables ▶

$$\frac{\partial^2 \Theta}{\partial x^2} = X''T \quad \text{and} \quad \frac{\partial \Theta}{\partial t} = XT'.$$

Equation (43) becomes

$$XT' = \alpha X''T - \gamma XT,$$

and dividing by XT gives

$$\frac{T'}{T} = \alpha\frac{X''}{X} - \gamma.$$

This can be rearranged as

$$\frac{1}{\alpha}\left(\frac{T'}{T} + \gamma\right) = \frac{X''}{X}.$$

Again, a function of x is equal to a function of t, so both must be constant. Choosing the constant to be $\mu = -k^2$, as before, the equations become

$$X'' + k^2 X = 0 \quad \text{and} \quad T' + (\alpha k^2 + \gamma)T = 0.$$

The boundary conditions reduce to

$$X(0) = 0 \quad \text{and} \quad X(L) = 0.$$

Solving the differential equation for $X(x)$ subject to fixed endpoint boundary conditions leads again to $k = n\pi/L$, for any positive integer n, and then to the family of solutions for X given by

$$X(x) = B \sin\left(\frac{n\pi x}{L}\right), \quad n = 1, 2, 3, \ldots,$$

where B is a constant.

The solution for T is given by equation (8) with $k = n\pi/L$:

$$T(t) = A \exp\left(-\left(\frac{\alpha n^2 \pi^2}{L^2} + \gamma\right)t\right), \quad n = 1, 2, 3, \ldots,$$

where A is a constant.

So the family of solutions is

$$\Theta_n(x, t) = a_n \exp\left(-\left(\frac{\alpha n^2 \pi^2}{L^2} + \gamma\right)t\right) \sin\left(\frac{n\pi x}{L}\right), \quad n = 1, 2, 3, \ldots,$$

where the $a_n = BA$ are constants.

Now use the principle of superposition to write down a more general solution:

$$\Theta(x, t) = \sum_{n=1}^{\infty} a_n \exp\left(-\left(\frac{\alpha n^2 \pi^2}{L^2} + \gamma\right)t\right) \sin\left(\frac{n\pi x}{L}\right).$$

Setting $t = 0$ gives

$$\sin\left(\frac{\pi x}{L}\right) = \Theta(x, 0) = \sum_{n=1}^{\infty} a_n \sin\left(\frac{n\pi x}{L}\right).$$

By inspection, $a_1 = 1$, and $a_n = 0$ for $n = 2, 3, \ldots$, as the term on the left-hand side is one member of the sum on the right-hand side. Therefore the solution is

$$\Theta(x, t) = \exp\left(-\left(\frac{\alpha \pi^2}{L^2} + \gamma\right)t\right) \sin\left(\frac{\pi x}{L}\right)$$

$$= e^{-\gamma t} \exp\left(-\frac{\alpha \pi^2 t}{L^2}\right) \sin\left(\frac{\pi x}{L}\right).$$

This is the same as the solution for the insulated rod except for the $e^{-\gamma t}$ factor, so the uninsulated rod cools more quickly than the insulated rod, by a factor of $e^{-\gamma t}$.

Index